Puzzles for Pleasure

Puzzles for Pleasure

Barry R. Clarke

CAMBRIDGE
UNIVERSITY PRESS

Published by the Press Syndicate of the University of Cambridge
The Pitt Building, Trumpington Street, Cambridge CB2 1RP
40 West 20th Street, New York, NY 10011-4211, USA
10 Stamford Road, Oakleigh, Melbourne 3166, Australia

First published 1994

A catalogue record for this book is available from the British Library

Library of Congress cataloguing in publication data available

ISBN 0 521 46634 2 paperback

Transferred to digital printing 2004

This book is dedicated to my family

SE

Contents

Advanced puzzles   55

Foreword

I first became acquainted with Barry R. Clarke's work when he joined me among the Brain Twister setters at *The Daily Telegraph* in 1989 and we later met when *The Daily Telegraph Book of Brain Twisters* was being planned.

Barry has brought an interesting and unusual background into the field of problem setting – though problem setters always tend to be a bit unusual. He was bored with school and spent his time inventing games. He left school at 16 and drew cartoon strips – a talent well displayed in this book. He then became an engineering apprentice, eventually gaining a place at Hull University where he read physics, concentrating on original papers and encountering Kant, and did an MSc in quantum mechanics. Finding his PhD program at Swansea uncongenial, he left and went to work as a copy editor, then as a financial software programmer in the City, with sidelines of writing comedy sketches for the BBC (*Alas Smith and Jones, Little and Large*) and busking in the London Underground. He retains an interest in physics and philosophy, harbouring a desire to solve some of the great puzzles presented by them.

Each problem creator has his own favourite problem types and criteria of excellence. Barry says his primary aim is to entertain, so he likes to create a novel and funny situation that entices the reader to carry on and then get stuck into the problem. The problems in this book clearly illustrate his success in achieving his aim. His favourite invention is the digital deletion problem, and he gives several examples here – e.g. pp. 15, 29, 40. He also likes to create more complex, carefully crafted puzzles, sometimes with part of the necessary information skilfully, even sneakily, concealed. The Advanced puzzles section provides a number of these that will be a trial for the hardiest of solvers.

I am especially pleased that Barry has included an outline history of recreational mathematics. Too often we forget that mathematics has a long history. It is one of the principal creations of the human mind and its history is as significant and as worthy of study as the history of religion. The history of recreational mathematics given here provides a readily comprehensible synopsis of the history of mathematics and many fascinating glimpses into cultural history.

David Singmaster
South Bank University
London

Acknowledgements

INFORMATION
The Bodleian Library
The British Library
Rhodes House
Professor David Singmaster

INSPIRATION
Peter and Karen Bertie
Rex Bradley
Richard Lea

OPPORTUNITY
Bernice Davison
Val Gilbert
David Tranah

PERMISSION
The Sunday Times

PUZZLE TESTING
Rejuan Ahmed
Andrew and Sue Varney

I am grateful to Andrew Varney for his valuable comments at proof stage.
Finally, the responsibility for any errors is entirely mine.

Preface

If you enjoy thinking, this book is definitely for you! Have a quick browse through the pages. There are brain twisters for all types of mind: logic problems, word conundrums, algebraic teasers, visio-spatial puzzles, lateral mind benders – in fact, no mind is left untwisted! You can forget about enrolling for a Ph.D. in Complex Number Theory to attempt this book. No puzzle requires a greater knowledge of mathematics than high school algebra and most don't even need that. The puzzles are divided into two levels of difficulty: Popular Puzzles which yield their secrets with a modest degree of effort, and Advanced Puzzles which demand much greater insight and concentration. The latter are intended for those puzzle geniuses amongst you who find the Popular Puzzles too easy. If you manage to solve at least half of the Advanced Puzzles then award yourself the title Professor of Pleasurable Puzzles!

Around half of the puzzles have been published before, mainly as Brain Twisters in the Saturday edition of *The Daily Telegraph*, while many are entirely new. All the puzzles are original in conception although puzzle enthusiasts will recognize Wire Wizards and In The Same Boat as developments of existing problems.

Some puzzles appear for special reasons. Although Find The Burglar follows a well-worked format, it was one of the first puzzles I ever composed, written when I was just 15 years old. A Tall Story appears as my first published mathematical puzzle from March 1987, while Save The City and Sum Secret are examples from the class of Digital Deletion Sums that I discovered in June 1987. The Word Bandit is a new word puzzle based on the principle of the one-armed bandit. Here, letters replace the traditional fruit symbols, and puzzle clues are solved to discover the positions where nudges occur, thus making a word.

For me, a puzzle has to be something more than a dry mathematical problem. I want it to be fun. Drawing on my experience as a TV comedy writer and newspaper cartoonist, I have tried to capture a sense of recreation in each problem. My wish for you is that if you get stuck on the puzzle then you at least get some pleasure from the illustration.

So there we have it! It's not impossible to solve all the puzzles in the book. *I* did (but then that's the advantage of compiling them!). Happy puzzling!

Barry R. Clarke

September 1993
Oxford

A history of recreational mathematics

The roots of recreational mathematics are inextricably tied up with the origins of mathematics itself. Their methods are the same, and as we turn back the clock to meet our ancestors we shall see several examples of a mathematical method expounded in a recreational context. There are also several surprises in store when we see the level of advancement of our predecessors. It turns out that prehistoric man was more than just a gibbering tree swinger; he actually had a coherent system of counting and, as far back as 4000 years ago, the Babylonians were not only calculating with numbers but actually had algorithms for solving quadratic equations.

Prehistoric man

Around 30 000 years ago, the natives of central Czechoslovakia had a base 5 system of numeration. This became apparent in 1937 when Karl Absolom reported the discovery of a prehistoric wolf bone with 55 cut notches arranged in groups of five, the first 25 being separated from the rest by a double-length notch (Bunt, Jones and Bedient 1988, p.2).

The real birth of mathematics, and with it recreational mathematics, probably began about 10 000 years ago with the Agricultural Revolution in the big river valleys of Mesopotamia (Beckmann 1971, p.18). With the growth of the farming culture came new problems demanding accurate forecasts of the seasons and better surveying techniques. This gave rise to a growth in astronomical and geometrical research. Although there are no known examples of mathematical thinking from this period, there is evidence of a reliable counting system. Around 8000 BC, in the Near and Middle East, there was a counting device in operation using the principle of the abacus (Fauvel and Gray 1987, p.44).

Neolithic Age (3000–2500 BC)

The Neolithic Age saw the development of written-number mathematics. The earliest known example is from the ceremonial mace of the Egyptian king Menes who lived around 3000 BC (Bunt *et al.* 1988, p.1). The mace claims the capture of 400 000 oxen, 1 422 000 goats and 120 000 prisoners, recorded in the form of coded numbers and illustrated livestock. Whether or not Menes ever suffered from insomnia is unclear, but if he did, his counting prowess probably cured it!

Around the same time, the natives of eastern Scotland were exploring geometrical objects. Examples of rounded regular polyhedra have been found carved into stone balls (Singmaster 1993*a*, p.1). Since they have no obvious application, they were probably recreational. Just as astonishing is the investigation of large stone remains in southern England and Scotland by Alexander Thom (Fauvel and Gray 1987, pp. 8–9). Thom conducted a statistical analysis of the measurements of hundreds of perimeters of neolithic rings and concluded that the measurements were an integral multiple of a standard unit, the 'megalithic yard' (MY), measuring 2.72 feet. Exactly 40 out of 163 sample items supported his conclusion to within ±0.1 MY. Thom also concluded the existence of the Pythagorean triplets (integer solutions to Pythagoras's theorem) 3,4,5 and 12,35,37 in his data thus supporting his 'megalithic yard' idea. If such a standard unit existed and it was stored somewhere, then it appears that the natives of southern England and Scotland were in closer communication than was first thought.

While the northern Europeans were discovering Pythagorean triplets, the Sumerians, in southern Mesopotamia, were calculating with large numbers. Jestin No. 50, a Sumerian text dated at around 2500 BC, probably gives the first example of a mathematical problem (Fauvel and Gray 1987, p.42). The large number of men involved (164 571) and the large number of sila (1 152 000) indicates that the problem was recreational. More significant is the fact that 7 is the first positive integer that produces a remainder when divided into 1 152 000.

> The grain [is] 1 silo. 7 sila each man received. Its men: 45,42,51. [How many sila are left over? The answer is] 3 sila remaining.

The first piece of information required to do the problem is that the Sumerians counted in the sexagesimal system. This means that the number x,y,z can be written as $60^2x + 60y + z$. Secondly, 1 silo = 40,0 gur and 1 gur = 8,0 sila. So in decimal notation, 1 silo = 1 152 000 sila. We now only need to multiply 164 571 men by 7 sila to arrive at 1 151 997 sila with 3 sila left over.

The early second millennium BC

The Rhind papyrus, was named after the Scottish antiquary A. Henry Rhind who bought the manuscript at Luxor, a Nile resort town, in 1858 (Bunt *et al.* 1988, pp.5–6). Now resident in the British Museum, it is the oldest known Egyptian mathematical document in existence. It was originally found in a ruined building at Thebes, and was written by an Egyptian scribe called Ahmes in 1650 BC, who writes (Beckmann 1971, p.21) that he copied the book

> . . . in likeness to writings made of old in the time of the King of Upper and Lower Egypt Ne-mat'et-Re.

This places the date of the original text between 2000–1800 BC. The document contains 84 problems and solutions, usually without the calculation. Problem 50 gives the value of π in use at the time which was 256/81 = 3.160 49. . . (Beckmann 1971, p.22). Problem 79 relates to a geometrical progression (Bunt *et al.* 1988, p.39):

> Sum the geometric progression of five terms, of which the first term is 7 and the multiplier is 7.

The given solution suddenly appears in a recreational setting:

houses	7
cats	49
mice	343

spelt	2401	[ears]
hekat	16807	[of grain]
	19607	

It is noteworthy that Leonardo of Pisa (Fibonacci) published his book *Liber Abaci* in 1202 with a rhyme which began 'As I was going to St. Ives . . .' based on the same principle (Wells 1992, p.3), though it is unlikely that he had access to the Rhind papyrus at the time.

The Old Babylonian period (1800–1600 BC)

The Old Babylonian period produced several examples of algebraic thinking. The Babylonians used a stylus to press their cuneiform (wedge-shaped) script into soft clay tablets which were then left out in the sun to harden. One such tablet, known as YBC 6967 (Fauvel and Gray 1987, pp.28–9), gives an early example of the solution to a quadratic equation.

[The igib]ūm exceeded the igūm by 7. What are [the igūm] and igibūm? . . . What the solver needs to know is that in this particular problem, the terms 'igibūm' and 'igūm' implicitly mean that their product is 60. This then leads to the two equations

$$x - y = 7$$
$$xy = 60$$

which give the quadratic equation

$$x^2 - 7x - 60 = 0$$

Although these equations were not explicitly written down, an algorithm was given to find the solution which is equivalent to the following calculation:

$$x = \sqrt{((7/2)^2 + 60)} + (7/2) = 12$$
$$y = \sqrt{((7/2)^2 + 60)} - (7/2) = 5$$

Note that the possible negative value for the square root was not given, missing the solution $x = -5$, $y = -12$.

A text, whose real purpose forms a puzzle in itself, is Plimpton 322 which now resides in the George A. Plimpton Collection, Rare Book and Manuscript Library, Columbia University (Fauvel and Gray 1987, p.33).

c^2/b^2	c	a
0.9834	119	169
0.9492	3367	4825
0.9188	4601	6649
0.8863	12709	18541
0.8150	65	97
0.7852	319	481
0.7200	2291	3541
0.6927	799	1249
0.6427	481	769
0.5861	4961	8161
0.5625	45	75
0.4894	1679	2929
0.4500	161	289
0.4302	1771	3229
0.3872	56	106

Figure 1. Plimpton 322, corrected and in decimal notation.

Dated at around 1600 BC, the tablet (Figure 1) shows three columns and 15 lines of sexagesimal numbers. The second and third columns form a list of integer pairs (c,a) for which there is an integer b satisfying (Stillwell 1989, p.3)

$$a^2 = b^2 + c^2$$

The first column gives the square of the gradient c^2/b^2 for these Pythagorean triplets, listed in descending order. Several features of these numbers are worth pointing out. Calculation of the angles associated with the gradients gives a range of 44.76–31.89 degrees with varying increments. There appears to be no pattern in these angles except for the fact that they are in descending order. Furthermore, there are written mistakes for some c but the corresponding gradient squared is correct. This suggests that either the text was copied or the calculations rested on unrecorded generating integers (p,q) which were related to a,b,c by (Fauvel and Gray 1987, p.38)

$$a = p^2 + q^2$$
$$b = p^2 - q^2$$
$$c = 2pq$$

This would make c and the gradient squared mutually independent. If this assumption of the pairs (p,q) is correct, the Babylonians not only knew some Pythagorean triplets but also how to find them. The fact that the text has 15 lines of independent Pythagorean triplets seems to confirm this. This meant that they had Pythagoras's theorem before Pythagoras (580–497 BC).

It is worth emphasising that the first column is the gradient squared. If the generating integers had been used, and the gradient had been required, it would have been easy to get it from $2pq/(p^2 - q^2)$. In contrast, the calculation of the gradient from a and c first involves finding the gradient squared $c^2/(a^2 - c^2)$ before taking the square root. Either way, the mathematician chose not to find the gradient but its square. This appears to rule out its use as a surveying aid and the purpose of the tablet remains a mystery.

The first millennium BC

Pythagoras is popularly credited with the right-angled triangle theorem that bears his name, a theorem that is often the basis of many geometrical puzzles. However, it is possible he was given more credit than he was due. Born around 580 BC at Samos near Turkey, he learnt mathematics from Thales then settled in Croton at the age of 40 (Stillwell 1989, pp.11–12). The school he founded there had a strict code of secrecy and all discoveries made there became its property. It is easily possible that one of his followers discovered the theorem and was sworn to silence. In fact, there is evidence that Pythagoras had a rather doubtful character. In an attempt to extend his influence to the masses he entered politics, and he became so unpopular that in 497 BC he was murdered.

Archimedes deserves attention as a creator and solver of several difficult problems, both recreational and practical. He was the greatest mathematician of ancient times and was the first writer to synthesise mathematics and physics. Born around 287 BC in Syracuse, son of the astronomer Pheidias

(Beckmann 1971, Ch. 6), he studied at the University of Alexandria under Euclid's successors. His friendship with Heiron II, the king of Syracuse, gave rise to the famous 'Eureka' incident, where the king asked Archimedes to confirm his suspicion that his crown was not pure gold. Archimedes found the answer as he observed the water level rising on climbing into his bath tub. The displacement of water on submerging the crown would have allowed him to find its volume and a measurement of its weight would have then allowed him to find its density which he could compare with the known density of gold.

He also developed the idea of limits and found the first accurate method of determining π. Using perimeter calculations of an n-sided regular hexagon inscribing a circle and one circumscribing the same circle, he placed the circle circumference and hence π between two limits. Then by successively doubling the number of sides of the polygons, he obtained, for n=96

$$3(10/71) < \pi < 3(1/7)$$

There are also several recreational problems that bear his name. His Cattle Problem (Wells 1992, pp.8–9), which was dedicated to the famous astronomer Eratosthenes, has a solution with more than 200 000 digits. It seems that Archimedes was quite comfortable with gigantic numbers. In his book, *Psammites* (*The Sand Reckoner*), he estimated that the number of grains of sand in a sphere the size of the then accepted universe was 10 to the power of 51 (Boyer and Merzbach 1989, Ch. 8).

Unfortunately, Archimedes did not have a happy ending. During the Second Punic War, the city of Syracuse was beseiged by the Romans and Archimedes helped to defend it. Despite his ingenuity at designing stone-hurling catapults and devices to set fire to the Roman ships, the city fell and Archimedes was slain by a Roman soldier, against the orders of the Roman general Marcellus. He was 75 years old.

The Chinese already knew their Gōugǔ theorem (Pythagoras's theorem) in the first century BC (Li and Du 1987, Ch. 2). This appeared explicitly in an astronomical book called the *Zhōubì suanjīng* at this time and probably dates from much earlier. The earliest mathematical treatise to be found is the *Jiǔzhāng suànshù* (*The Nine Chapters on the Mathematical Art*), which was rewritten and augmented several times, mainly by Marquis Zhāng (d.152

BC) and Gĕng Shòuchāng (73-49 BC), and represents the accumulated knowledge of Chinese mathematics from the eleventh century BC to AD 220. In the discussion of fractions, the first use of the least common multiple appears, predating its discovery by Leonardo of Pisa (Fibonacci) in the thirteenth century AD. The work also introduces into mathematics positive and negative numbers, which the Indian mathematician Brahmegupta later found around AD 620. An outstanding feature of the Chinese text is the solution of linear simultaneous equations using rectangular arrays. Problem 1 in the section 'Rectangular arrays' reads

> Top-grade ears of rice three bundles, medium-grade ears of rice two bundles, low-grade ears of rice one bundle, makes 39 dou; top-grade ears of rice two bundles, medium-grade ears of rice three bundles, low-grade ears of rice one bundle makes 34 dou; top-grade ears of rice one bundle, medium-grade ears of rice two bundles, low-grade ears of rice three bundles, makes 26 dou. How many dou are there in a bundle of top-grade, medium-grade, low-grade ears of rice?'

In present-day algebra, this problem would be formulated as

$$3x + 2y + z = 39$$
$$2x + 3y + z = 34$$
$$x + 2y + 3z = 26$$

The Chinese solved this by placing the coefficients in a rectangular array and subtracting multiplied rows, one from another, to produce zeros. The earliest discovery of this method in European civilisation was by Buteo in France in the sixteenth century AD. This puts Chinese problem solving in the first millenium BC far ahead of contemporary European development.

The Dark Ages

In Europe in the late eighth century, the river crossing problem made its first appearance. It came from the pen of Alcuin, an English scholar, who recorded (Hadley and Singmaster 1992)

> A man had to take a wolf, a goat, and a bunch of cabbages across a river. The only boat he could find could only take two of them at a time. But he had been

ordered to transfer all of these to the other side in good condition. How could this be done? Since wolves eat goats and goats eat cabbages, neither pair could be left alone. However, there are two solutions to the problem. The goat must go across first, but then either the wolf or the cabbage can be taken across next before returning with the goat.

Alcuin was born near York around AD 732 and spent much of his student and working life at the cathedral school in York. From 781–96 he was educational advisor to Charlemagne, the greatest military and political ruler of the Dark Ages. In Aachen, under Charlemagne, Alcuin directed the major reform of learning in Europe.

Alcuin's main work, *Problems to Sharpen the Young*, from which the river-crossing puzzle (problem 18) is taken, is the earliest known collection of problems in Latin. It contains 56 problems, including the rather amusing slug problem

A leech invited a slug for lunch a leuca away. But he could only walk an inch a day. How many days will he have to walk for his meal?

The problem has the unfortunate answer that since a leuca is 90 000 inches, the slug takes 246 years and 210 days. No doubt the slug had thoroughly deserved his lunch by the time he arrived!

The second millennium AD

In AD 1175, Leonardo of Pisa (Fibonacci) was born. He played an important part in introducing Indian numerals to Europe, which he learnt from Arabs during his African visits. Fibonacci found fame with his discovery of the Fibonacci sequence (Wells 1992, p.27), which resulted from the following problem:

A certain man put a pair of rabbits in a place surrounded on all sides by walls. How many pairs of rabbits can be produced from that pair in a year if it is supposed that every month each pair begets a new pair which from the second month on becomes productive?

The classic 17 horses problem (Wells 1992, p.32) was presented by Niccolo Fontana (Tartaglia) in 1546

> A man dies leaving 17 horses to be divided amongst his heirs in the proportions 1/2:1/3:1/9. How can this be done?

Fontana's solution involved borrowing an extra horse to calculate the distribution, politely returning it after the calculation. In fact, one need only multiply the proportions by 18 to arrive at the solution 9,6,2.

In the nineteenth century, two important figures stand out: the Englishman Henry Dudeney and the American Sam Loyd. Both were prolific compilers, though Dudeney is widely regarded as the better mathematician while Loyd is seen as the better puzzles promoter. Dudeney did pioneering work in digital roots, realising that one could verify that a number is not a square by repeatedly summing digits until a single digit remains. If the answer is not 1,4,7 or 9 then the number is not a square. For example, 43 414 922 is not a square because the successive sums, 29 then 11, reduce to 2. He was also a master of dissection problems, one of his more famous ones requiring the cutting of an equilateral triangle into four pieces so that they could be reassembled into a square (Newing 1988–9).

Sam Loyd was undoubtedly a great chess problemist and a prolific composer of mathematical puzzles. However, a degree of caution is needed when assessing his contribution because several of his claims to priority were unjustified. He is often credited with the invention of the cryptarithm or alphametic, where the digits in an arithmetic calculation have been replaced by letters, the aim being to recover the digits. However, an example appears in *The American Agriculturist* of 1864 (Singmaster 1993*b*), and since the 23-year-old Sam Loyd was still preoccupied with chess problems at the time, this example almost certainly predates his contributions. Some doubt is usually expressed about what Loyd actually claimed to have invented. His headed notepaper (Figure 2) dated 15 April 1903 settles the issue, clearly stating 'Author of the famous "Get Off The Earth Mystery", "Trick Donkeys", "15 Block Puzzle", "Pigs In Clover", "Parcheesi", Etc., Etc.,' (White 1914, Plate III). In fact, the "Get Off The Earth Puzzle" was a circular version of linear disappearing objects puzzles (Gardner 1956, Chs. 7–8) and the principle of the "Trick Donkeys" puzzle had already been

SAM LOYD,

Journalist and Advertising Expert,

ORIGINAL

Games, Novelties, Supplements, Souvenirs,
Etc., for Newspapers.

Unique Sketches, Novelties, Puzzles,&c.,
FOR ADVERTISING PURPOSES.

Author of the famous
" Get Off The Earth Mystery," " Trick Donkeys,"
" 15 Block Puzzle," " Pigs In Clover,"
" Parcheesi," Etc., Etc.,

P. O. BOX 826.

New York, *April 15* 190 *3*

Figure 2. Sam Loyd's headed notepaper.

used with dogs (Singmaster 1993*a*, p.222) around 1857, thirteen years
before Loyd registered it.

As for the "15 Block Puzzle", there are strong reasons for entertaining
doubt as to Loyd's priority. Fifteen numbered blocks were randomly placed
in a 4 × 4 grid with one vacant square. The aim was to use the vacant space
to slide the blocks into serial order, leaving the space in the bottom right-
hand corner of the grid. It was the greatest puzzle craze of the nineteenth
century and occupied most of America and Europe from 1879 to 1881. The
New York Times reported twice only on the craze, on 22 March 1880 and
again on 11 June 1880, and although Loyd actually lived in New York, he
curiously failed to receive credit. Other contemporary articles on the craze
have also been found (Hordern 1986, p.20) and again Loyd is not mentioned.
Given the master self-promoter that Sam Loyd was, his absence from

publicity is uncharacteristic, especially since the puzzle was a world-wide craze.

The "14–15" puzzle was a special initial configuration of the "15" puzzle, arrived at by placing all the blocks in serial order except the 14 and 15 which were juxtaposed. Loyd put up a prize of $1000, indicating that it was his own invention. In fact, the inventor could have been just about anyone who attempted to solve the "15" puzzle. For example, reporting on the "15" puzzle on 22 March 1880, the *New York Times* revealed that

> ... At 8 o'clock the next morning Mr Schurz was taken home in a carriage, completely exhausted, and leaving his blocks in the position 13,15,14 . . . .

Whoever reached this configuration first, Sam Loyd's son (Loyd 1928, p.1) did not think it was his father:

> It was in the early 80s, when I had barely attained my 'teens, that the "14–15" puzzle flashed across the horizon, and the Loyds were among its earliest victims.

So the "14–15" puzzle found Sam Loyd, not vice versa. To this day, the real authors of the "15" puzzle and its insoluble derivative are still unknown.

In the spring of 1974, an architect from Hungary conceived a magic cube for demonstrating spatial moves (Rubik *et al.* 1987). His name was Ernö Rubik and the cube became known universally as Rubik's cube. Although marketed in Hungary in 1977, it took until 1980 for the cube to catch on. In 1980, a million cubes were sold in Hungary alone, amounting to purchases by one tenth of the population. In the same year, the American Ideal Toy Company, who exported the cube out of Hungary, estimated their sales to be 10 million. Many awards followed, and it became the only game to win the UK Toy of the Year award twice, first in 1980 and again in 1982. The cube was so popular that in 1982 a World Championship was set up. Budapest hosted the event with contestants competing from 19 countries and the US champion Minh Thai clinched the title with a time of 22.95 seconds. The craze was not to last. Unfortunately, cheap copies from the Far East started to flood the market and retailers lost interest. Consequently, in 1983, the greatest puzzle craze of the century began to die out.

It's interesting to imagine how Sam Loyd would have exploited his puzzles with the power of today's media. He might be hosting some peak-time puzzle show on TV, with six-figure dollar prizes. And with numbers as big as that, Archimedes would probably be helping him!

Popular puzzles

Wong sum

'Today we shall do an exercise in subtraction,' said Mr Adder as he scribbled an equation on the blackboard.

Taking the board rubber, he rubbed out a digit on the right-hand side of the equation.

'Now then, what must I subtract from the left-hand side to make both sides equal?'

Wong, the new Chinese pupil, rose from his seat, took the rubber and erased a digit on the left-hand side. 'Well, that makes both sides equal,' chuckled Adder, 'but it's not quite what I meant. Have another try.' With that, Adder rubbed out a second digit on the right-hand side. Wong studied the problem for a moment then again erased a digit on the left-hand side to balance the equation.

'Er, you still haven't got the hang of it, have you?' said Adder with a hint of desperation. 'One final go.' With that he rubbed out one of the remaining two digits on the right. To Adder's astonishment, Wong again erased on the left to make both sides equal.

'Very interesting,' remarked Adder, scratching his head. 'What kind of subtraction is that?' Wong smiled. 'It's called Chinese take-away!'

If all gaps left by erased digits were considered to be closed up, what were the three equations created?

Strange street

Mrs Gossip was telling her friends the latest. The woman at 5 or 9 had run off with the milkman; the couple at 5, 7 or 11 were holding a pyjama party, but without the pyjamas; the skinhead at 5, 7 or 9 had assaulted the vicar; and the hippy at 9 or 11 was high again (in fact, he was sitting on the roof). They live in separate houses and the couple live next door to the hippy.

What numbers do they occupy?

Hints p.79   *Solutions p.92*

Harvey's wall climber

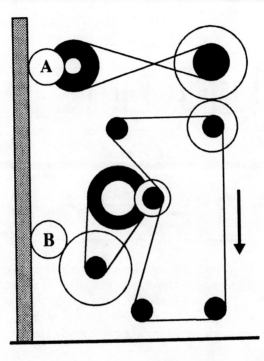

When Harvey Screwbrain invented his wall-climbing machine, everyone thought it wouldn't work. They were right! Quite simply, there was nothing to hold the machine onto the wall. But how much of a screwbrain was Harvey? Did he at least manage to get the wheels that were in contact with the wall (A and B) to turn the right way? The inside of the machine is shown above with pulleys (black) and cogs (white).

Which way do wheels A and B rotate?

Hints p.82 *Solutions p.94*

Word bandit 1

The letters in the Word bandit are numbered from 0 to 9 and conceal a 10-letter word. To find the word, each letter can either be retained or nudged (moved) up to two places forwards or backwards in the alphabet. The moves for some of the letters can be found from the puzzles below, where each solution is a digit revealing the position of the letter to be moved.

Nudge 2 forwards

'My Dad is three times my age,' declared Dimple, 'but he's the square of my younger brother's age. I'm twice as old as my younger brother.'
What age was his younger brother?'

Nudge 1 backwards

Weedy Willie's vegetable garden was surrounded by a fence in the shape of a regular polygon, with a fence post at each corner. To keep the birds away, each pair of non-adjacent posts was joined by a length of silver string stretched across the garden. There were 20 lengths of string. How many sides did the polygon have?

No nudges

'An interesting digit,' remarked Professor Brainbloom. 'The number is the same as its number of letters when spelt out.'
What digit did Brainbloom have in mind?

Nudge I backwards

$$62 ? x = 29 ! x$$

Excavations in Lower Cranium once unearthed a clay tablet with an arithmetic equation engraved on it. Unfortunately, time had taken its toll and only two of the numbers and the equals sign remained. The calculation is shown above where ? and ! represent different arithmetic signs and x is the same digit.
What is the missing digit?

Nudge I backwards

Toddler Toby was building a gigantic cube from his cubic building bricks. Clearly, the 5400 bricks in his mega toy cupboard were not enough. What is the lowest digit by which 5400 must be multiplied to arrive at a number of bricks from which Toddler Toby can build a cube?

Solutions p.89

𝒫rime of life

'I've always been 45 years older than your dad,' said Grandma to young Trickle. Trickle always suspected that Grandma was a bit short on grey matter but now her statement of the obvious really clinched it.

'But I'll tell you what's strange about our ages now,' she continued. 'The two digits in my age are the reverse of the digits in your dad's age. And what's more, they're both prime digits.'

Trickle couldn't believe his ears. He'd thought Grandma was as daft as a carrot and here she was making mathematical observations. Trickle felt ashamed as he'd often joked about Grandma's brains behind her back. Mmm, maybe that's where she'd been hiding them all these years!

How old is Grandma?

Hints p.81   *Solutions p.96*

The horse and the hurdle

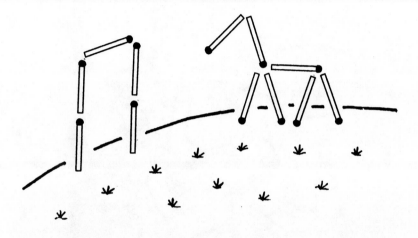

As Hobble the horse stood on the hillside staring at the hurdle below, he wondered if he would ever have the courage and strength to jump it. He dreamt that one day, he would be facing the hurdle from the hill on the other side, an equal distance away. Move exactly six matches so that this is so.

Cornflakes and porridge

In Duff Row, there are seven prison cells in a line. These are numbered from one to seven from left to right, exactly six being occupied. The vacant cell awaits the arrival of Cornflake Colin, the serial killer whose victims were force-fed until they flaked out.

In the occupied cells, Poisonous Pat was next to Harry the Hatchet, Gruesome Gertie was next to Slasher Sam who was two to the right of Desperate Deborah. Harry the Hatchet was three away from Vicious Vince. Desperate Deborah was two away from Poisonous Pat.

Who will be Cornflake Colin's two neighbours?

Hints p.79   *Solutions p.92*

Railway rhyme

Platform 1 we departed at nine,
'Au Revoir!' to Beaujolais wine,
Reminisces to toast,
In our train to the coast,
State the place at the start of the line.

Safe conduct

Bumbletown had the most robbed bank in the land. The unfortunate clerk was frequently forced to open the safe, and the bank had lost so much money, that Mr Good, the bank manager, was going bald.

Then one day, Mr Good had an idea. His nephew, Fumble, should be the bank clerk. Now Fumble was the ideal man for the job. His memory was so bad, one could be sure that no robber could ever force him to remember the safe combination. Furthermore, his poor powers of recall were matched by a superb talent for puzzling things out. This meant that whenever Fumble needed to know the safe combination, all he had to do was obtain the following conundrum from Mr Good, which he could solve to reveal the five-digit safe combination.

'The fourth digit is four greater than the second digit. There are three pairs of digits that each sum to 11. The third of the five digits is three less than the second. The first digit is three times the fifth digit.'

Of the 100 000 possible numbers, which was the correct safe combination?

Hints p.81   *Solutions p.89*

The doubtful die

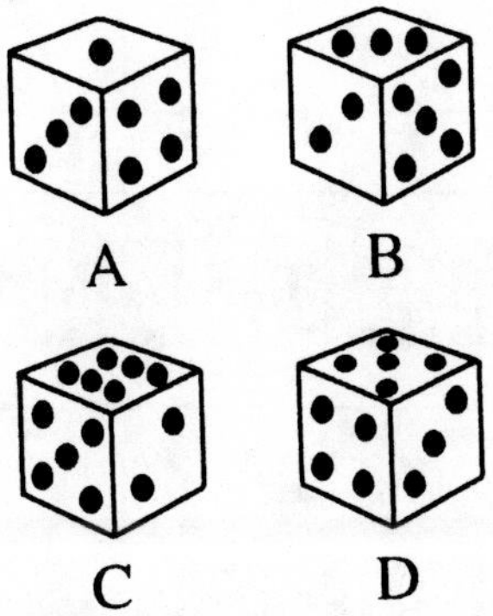

A   B

C   D

The Dopey Dice Company manufacture dice with opposite faces that do not all total seven, contrary to the case with normal dice. Not only that, but sometimes they make a die whose faces are orientated differently to those of their regular dice. This is the case in the diagram above, where three of the views are of the same die and the other view is of a rogue die.

Which is the odd one out?

Hints p.80   *Solutions p.92*

Word bandit 2

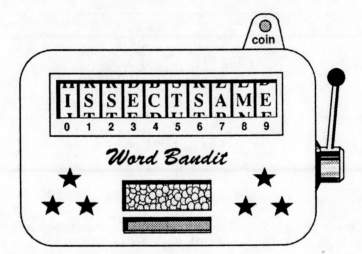

The letters in the Word Bandit are numbered from 0 to 9 and conceal a 10-letter word. To find the word, each letter can either be retained or nudged (moved) up to two places forwards or backwards in the alphabet. The moves for some of the letters can be found from the puzzles below, where each solution is a digit corresponding to the position of the letter to be moved.

Nudge 1 backwards

A shopkeeper placed three indistinguishable oranges in a square box with 3 × 3 compartments. How many ways can the oranges be placed in the box so that no two oranges appear in the same row or column?

Nudge 1 backwards

A square paper napkin has an area of 128 square centimetres and is folded in half along a line joining opposite corners. The triangular shape is folded in half again so that two corners meet and the folding is repeated leaving a triangular shape each time. Starting with the square napkin, how many times must it be folded to get the longest side of the triangle equal to 1 cm?

Nudge 2 forwards

Sleepy Sam had just woken up. If he had woken up at noon the previous day, he would be twice as many hours away from the present as the present is from noon. What hour is it?

Nudge 1 forwards

Professor Ponder had found a digit that appears in neither its square nor cube, and when multiplied by a digit that appears in both its square and cube, gives a product equal to one greater than the square of the first digit minus the square of the second. What digit had Ponder found?

Nudge 2 forwards

A backward robber walked into a drug store and said 'I want all the money in the safe minus the money in the cash register.' 'That's ten thousand dollars,' said the proprieter. The robber frowned. 'Then I'll take all the money in both,' said the robber. The store owner gave the robber seven thousand dollars, a half of the total requested. How many thousand dollars were in the cash register?

Solutions p.95

/Muddle market

At Muddle Market, a row of traders had their stalls ordered as follows: iron-monger, greengrocer, fishmonger, butcher, confectioner. Five elderly ladies from the Confused and Bewildered Club were out on a shopping trip. Now, each lady wanted to buy from one stall only, but no lady could remember from which one, though they remembered that no two of them wanted to visit the same stall. Mrs Folly wanted to buy from the ironmonger, green-grocer, fishmonger or confectioner; Miss Dippy from the fishmonger, butcher or confectioner; Mrs Grumble from the ironmonger, greengrocer or confectioner; Mrs Vacant from the ironmonger, fishmonger or confectioner; and Miss Witless from the ironmonger, butcher or confectioner. Miss Dippy and Mrs Grumble eventually bought from adjacent stalls. What a pity the two of them could not remember which ones, for then the others could have deduced the stalls that they intended to visit.

Can you give the trader that each lady bought from?

$\mathcal{F}$ractionally mean

At the reading of Elijah Polyp's will, his two sons Nabber and Grabber were eagerly waiting to learn how much land they had inherited. The big moment had arrived. The lawyer, who was rather drunk, fumbled in his briefcase, took out the will, and belched loudly.

'Out of the 8235 acres left to my two sons, Nabber gets 1647, and Grabber gets the rest.'

With that, the lawyer wrote the message 'Nabber 1647/8235' on his notepad and went in search of the toilet. Being mean, Grabber took the notepad and a rubber, and tried to reduce Nabber's share by rubbing out exactly one digit in the numerator and denominator. Curiously, the remaining six digits gave the same magnitude as before. So Grabber rubbed out a further digit on the top and bottom. Still the same magnitude! Footsteps in the corridor signalled the lawyer's return. In a last act of desperation, Grabber erased one last digit from the top and bottom. As the lawyer entered the room, Grabber realised that all his attempts had failed to alter the magnitude in front of him. The lawyer returned the notepad to his briefcase and Nabber and Grabber got their rightful proportions.

What was the order of the three pairs of digits that Grabber erased?

Hints p.80   *Solutions p.89*

Domino chain

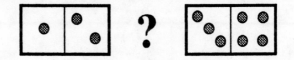

Down at the Pig and Bucket, the locals were supping ale and playing dominoes. Young Gibber, who was new to pub life, wanted to know the rules. 'Simple!' piped up Legless, one of the regulars. 'The rectangular tiles used are each divided into two squares. Each square carries a number of spots from 0 to 6, so that all the possible combination pairs appear once only in the set. Each player takes it in turns to place a tile so that a chain of dominoes is constructed. One of the numbers on the tile put down must match, and lie next to, the number at the end of the chain where it is placed.' This was a rare moment of eloquence from Legless.

Gibber looked at the domino chain on the table. He noticed that 25 tiles had been placed, the number of spots totalling 155. The end numbers of the chain were 2 and 3.

What dominoes remained to join the chain ends?

Signs of confusion

Livingstone Mortimer had been walking through the jungle for days. Suddenly, he came to Booliba village where he found a signpost which read 'Rumba 4, Wobble 7'. Heartened, he continued his journey. However, when he reached Rumba, he found a signpost showing 'Booliba 2, Wobble 3'.

Livingstone knew something was wrong, as the two signposts were clearly contradictory. However, he resumed his journey and soon reached Wobble. Here, the signposts read 'Rumba 4, Booliba 7'.

Livingstone was perplexed. He stopped an old man who was walking towards him and described the three inconsistent signposts.

'They're perfectly correct,' said the old man. 'At one of the three villages, the inhabitants are all honest, so their signpost is alright. At one of the villages they only tell the truth half of the time, so only one of the two numbers in their signpost is correct. The other village is full of liars, so neither of the numbers in their signpost is correct.'

If Livingstone followed a straight road, which inhabitants lived in which village?

Hints p.80   Solutions p.92

Room for assistance

At Muckrake Mansions, Inspector Twiggit was investigating a murder. The six suspects had joined him in the lounge for questioning. He knew that on the night of the murder Miss Lipstick was in the study, the kitchen or the dining room; Mr Britches was in the kitchen, the morning room or the dining room; Miss Uppity was in the study, the kitchen or the conservatory; Colonel Crumpet was in the kitchen or the morning room; Mr Splutter was in the study, the library, the morning room or the conservatory; and Professor Twinkle was in the kitchen, the library, the conservatory or the dining room. He also knew that each room had exactly one person in it.

The big moment had arrived.

'And now I shall announce who the murderer is,' said Twiggit. 'The murderer was . . . er, in the morning room.'

The trouble was, Twiggit had no idea who was in the morning room. If he had known, he could have uniquely determined everyone else's whereabouts.

Who was in each room?

Hints p.81   *Solutions p.96*

King-size conundrum

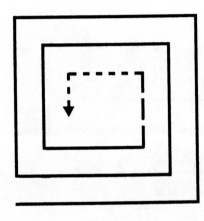

A medieval king needed to work out how he could recruit fighting men for the battle ahead. However, there were so many distractions around the castle, his thinking became confused. So, in order to change his daze into knights, he asked for a secluded walk to be made so that he could ponder in peace.

The head gardener was given the job of planting lines of high bushes. First, he planted a line running 100 paces east. Then from the end of that line he planted a line 100 paces north, then 100 west, 98 south, 98 east, 96 north, 96 west, and so on, dropping the measurement by two paces every second anticlockwise 90 degree turn. This made a square spiral path 2 paces wide.

If the king intended to walk down the middle of the path, how long was the path?

Word Bandit 3

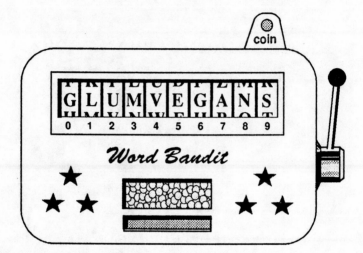

The letters in the Word bandit are numbered from 0 to 9 and conceal a 10-letter word. To find the word, each letter can either be retained or nudged (moved) up to two places forwards or backwards in the alphabet. The moves for some of the letters can be found from the puzzles below, where each solution is a digit corresponding to the position of the letter to be moved.

Nudge 2 forwards

The front wheel of Farmer Sprout's tractor has a radius of 25 cm and rotates at two and a half revolutions per second. The back wheel has a diameter of 125 cm. At how many revolutions per second does it turn?

Nudge 2 forwards

An apple, orange, banana and pear are laid out in a straight line. The orange is not at either end and is somewhere to the right of the banana. In how many ways can the fruit be laid out?

Nudge I forwards

Herbert's age is equal to twice the product of its two digits. How old was Emily when she was a quarter of Herbert's age now?

Nudge I backwards

Grovel the gardener had a circular garden with a regular hexagonal lawn placed so that all six corners were located on the garden perimeter. Now Sam knew that the lawn was made from $\sqrt{108}$ square metres of turf. What was the diameter of the garden (metres)?

Nudge 2 backwards

Mrs Frump had less than 15 plums in a bag. Grabbit took some, Rummage took less than Grabbit and Tumble took less than Rummage, so that no one had none and their total had two digits. If Grabbit and Rummage gave some to Tumble to make an equal distribution, how many had Rummage taken?

*Solutions p.*99

The six sheep pens

A fickle farmer had three sheep and three rams and wanted to erect an enclosure for them in his field. For this he enlisted his long-suffering son who was given 12 equal lengths of fencing and told to erect six pens, one for each animal.

The farmer decided that since a sheep was smaller than a ram he would have three small pens and three large pens. So he told his son to make six square pens so that each of the large pens was twice the area of each of the small ones. His son set to work and just as he had finished the farmer appeared.

Perhaps it would be better to have the large pens three times the area of the small ones. His son smiled and with a few adjustments produced the new ratio of areas. However, the farmer was still not satisfied. Maybe rectangular pens would be more suitable. His son was unperturbed, for he knew that he had found a single arrangement of the 12 fences, giving three small and three large pens, that could be adjusted to produce either square or rectangular pens and any ratio of areas.

How were the 12 fences arranged?

Hints p.81   *Solutions p.97*

$\mathcal{F}$ind the burglars

Five burglars, Jones, Smith, Peterson, Bloggs and Harper burgled five houses called Catalan, Belvedere, Oakmoor, Ardswick and Pinewood, but not necessarily in that order. The houses have jewellery, paintings, antiques, trophies and money in them but the order tells you nothing about which house contained what.

Exactly one burglar's initial corresponded with the initial of the goods he stole. The burglar who stole money had five letters in his name. Jones only stole things that had obviously been won. Precisely one burglar's initial corresponded with the initial of the house he burgled. Peterson burgled Catalan. Smith and Harper each burgled houses that had a tree in their names. Oakmoor had jewellery stolen from it.

Which burglar stole what from which house?

The enigmatic interview

The Pomphouse Poker Club had a curious entrance test. Four cards were dealt on the table by an assistant, one to each of the four interviewers. The eight faces on the four cards each had a colour. Two faces were red, two blue, two yellow and two green. This meant that four colours were visible and the other four colours were hidden. Each interviewer then looked at his hidden colour and made a statement about it.

Now on one occasion, the following statements were made. Mrs Globule said 'green or blue', Mr Crust stated 'neither green nor blue', Major Wilson-Bucket barked 'blue or yellow', and Miss Nostril claimed 'blue or yellow'. Their visible colours were, respectively, red, green, red and blue, and no card had the same colour on both faces. Exactly two of the interviewers were lying.

Can you work out what was each person's hidden colour and thus prove yourself a worthy applicant to the Pomphouse Poker Club?

Hints p.81   *Solutions p.90*

Sorry dad

After taking Mother to the cinema, I began to take her home. However, when I looked back I saw that I'd left Father.
What was his name?

Save the city

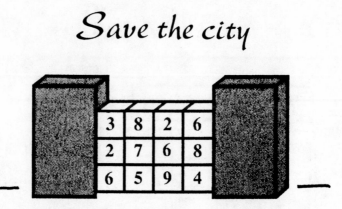

Near the ancient city of Citifon, the river was close to bursting its banks. Yants, the evil wizard, had wedged open the floodgates protecting the city using 12 wooden blocks, each bearing a digit, and arranged in a sum so that the top row added to the middle row gave the bottom one. Yants had declared that the gates could be closed only by removing the blocks in a definite sequence as follows.

Working from top to bottom, remove one block from each row to leave three columns of digits (the pressure of the gates closes the gaps), then a second to leave two columns, then a third to leave one column, and finally the last in each to close the gates, so that a valid sum remains each time.

What sequence saves the city?

Solutions p.99

Crumple Towers

Floor	1st name	2nd name	Rent day	Pets
4	Agatha	Cesspot	Tuesday	3
3	Sidney	Hump	Monday	2
2	Bertha	Bucket	Wednesday	4
1	Cuthbert	Nosebag	Friday	1

Reginald Scatter, the landlord of Crumple Towers, had written down the first name, second name, rent day and number of pets for his four tenants, with exactly one tenant on each floor. Each item had been written in the correct column, but only one item was correctly positioned in each column. Nevertheless, Reginald realised that he could work out the correct details from his recollections.

Bertha, who paid her rent on Tuesday or Friday, had neither 1 nor 4 pets in her flat.

Mr Nosebag paid rent on Tuesday or Friday and lived directly above the tenant with 4 pets, who was called neither Cesspot nor Bucket and who paid rent on Tuesday or Wednesday.

Miss Cesspot paid rent on Monday or Wednesday and had 2 or 4 pets. Can you give the first and second names for each floor together with the rent day and the number of pets?

Word bandit 4

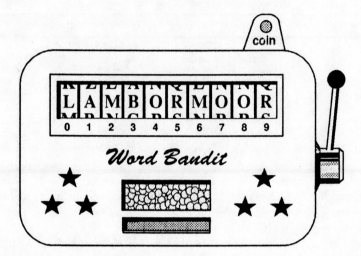

The letters in the Word bandit are numbered from 0 to 9 and conceal a 10-letter word. To find the word, each letter can either be retained or nudged (moved) up to two places forwards or backwards in the alphabet. The moves for some of the letters can be found from the puzzles below, where each solution is a digit corresponding to the position of the letter to be moved.

No nudges

Happy Harry's square birthday cake had been cut along the two diagonals. How many triangles of any size can be seen?

Nudge 1 backwards

Someone had rubbed out part of an equation in Tiny Tum's homework book. The remaining equation read

$$18 ? x = 45 ! x$$

where the ? and ! were two different arithmetic signs and x was the same digit on both sides. What was the digit?

No nudges

In a throw of three dice a two digit total was achieved. One die had a number equal to the sum of the two digits. The opposite faces of the other two dice corresponded to the two digits in the total. What was the sum of the two digits?

Nudge I forwards

Four schoolboys formed a line in the playground. Sidney wanted to stand next to his friend Ollie but at an end of the line. How many ways can the four schoolboys be arranged?

Nudge I forwards

A railroad runs due south from Gunslinger station to meet the Great Atlantic Railroad (GAR) at Smokeshot station. A second line runs due south-west from Gunslinger to meet the GAR at Deadgun station. A train sets out from Gunslinger to Smokeshot at the same time and speed as a train sets out due west from Smokeshot to Deadgun. If no train changes direction, how far is the second train from Deadgun when the first arrives at Smokeshot?

Solutions p.93

Romance on the stone

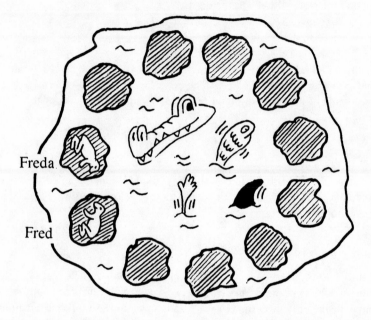

Frederick the frog quite liked Freda the frog who was sitting on the next stone to him on the pond. He began to wonder how many jumps it would take him to land on the same stone as her. The 11 stones were equally spaced in a circle around the pond. Frederick could jump over two stones at a time, landing three away, while Freda could clear one stone in each jump, landing two away. They both jumped simultaneously and Freda always jumped anti-clockwise.

In which direction should Frederick keep jumping for the quickest rendezvous, clockwise or anti-clockwise?

Quite a card

Auntie Sadie was tormenting her nephew Dribble. She had placed nine coloured cards face-down on the table in three rows and three columns and refused to let Dribble see them.

'If you want to know what they are,' snarled Sadie, 'you'll jolly well have to work them out.'

'But how?' asked Dribble, clutching his teddy bear. Auntie Sadie grinned maliciously.

'Well, the red card is in the first or second row. The third column has exactly two green cards. Exactly two blue cards are in the second row. Precisely three of the corners are occupied by yellow cards. In each row there is exactly one green card.' Dribble lost no time. In a flash, he and Teddy had worked them out.

How were the cards laid out?

Hints p.80
Solutions p.93

ᴄhe three piles of coins

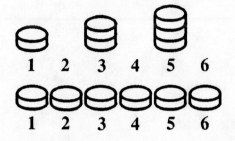

There were once six children who, while out playing in the local park, found six identical coins. In the scramble that followed, the first child took one coin, the third child took two coins, and the fifth child took three. When they got home, their father was so annoyed to learn that three of his children had been left empty-handed that he ordered the children to share out the coins so that they had one each. The situation is shown in the top diagram above where the six coins have been divided into piles of one, two, and three. Pick up exactly one coin once only, and use it to lay the six coins out in a line as shown in the bottom diagram. The chosen coin must not touch another coin while held, no other object can be used and the table cannot be tilted. N.B. large coins are best.

Cubic incapacity

The Threedee family had ordered some unit cubes from Twistem Ltd. The lorry arrived and the unit cubes were set out on the driveway in a one-layer square. The family assumed that the correct number had been delivered, because the ordered quantity, like the actual delivery, would make a one-layer square. However, Twistem had twisted 'em.

Now each member of the family wanted to construct a cube from the unit cubes so that there were 10 different sizes having sides from one to 10. Father and Mother were to build the biggest two cubes and the eight children would build the rest.

Nine members of the family took exactly the number of unit cubes they needed, but one of the twins, Puddle, was left with none.

How many units were missing?

$\mathcal{E}xtra$ $sensory$ $deception$

Psychic Sally claims to be able to 'sense' the nine different colours on a 3-by-3 grid of squares that a volunteer secretly fills in. Her assistant, Doubtful Doreen, who looks at the positions of the nine colours, makes five statements about their relative positions to Sally. What no-one in the audience realises is that it is possible to deduce the positions from the statements Doreen makes. On one occasion, the statements were as follows. (1) A red is directly above a blue. (2) A yellow is two to the right of a green. (3) An orange is two above a pink. (4) A turquoise is directly below a violet. (5) A white is directly to the right of a blue.

How were the colours arranged in the grid?

Hints p.81 *Solutions p.100*

True to the tribe

The Floppybottom tribe were choosing their new chief. This was to be the first tribesman who could correctly identify the honest members of the Grunter family, a problem that had long baffled the elders of the Floppybottoms.

The five Grunters were brought before the rest of the tribe. Of the five, two were known to consistently tell lies and the other three were known to be truthful, but no one had ever been able to decide which person did which.

So each Grunter was invited to make a statement about the other members of his family. Appu said 'Only one of Babble and Cowa tells the truth.' Babble claimed that 'None of Appu and Eva tells lies.' Cowa reported that 'None of Appu and Dobi tells lies.' Dobi said 'Only one of Cowa and Eva tells lies.' Eva claimed that 'Babble and Dobi are either both truthful or both liars.'

Which Grunters were honest?

Hints p.79　　　　　　　　　　　　　　　　*Solutions p.93*

Word Bandit 5

I	F	D	E	M	O	B	E	A	N
0	1	2	3	4	5	6	7	8	9

Word Bandit

The letters in the Word bandit are numbered from 0 to 9 and conceal a 10-letter word. To find the word, each letter can either be retained or nudged (moved) up to two places forwards or backwards in the alphabet. The moves for some of the letters can be found from the puzzles below, where each solution is a digit corresponding to the position of the letter to be moved.

Nudge 2 backwards

A coin with diameter 2 cm rolls around the curved edge of a coin with radius 8 cm until it reaches its starting point. How many complete revolutions does the small coin make?

Nudge 1 backwards

At a theatre, each row had 10 seats running from 0 to 9 from left to right.

Five friends occupied five consecutively numbered seats in a row. Barbara sat somewhere to the right of Deborah, Ernie was not next to Deborah, and Andrew sat two places to the left of Deborah. If Ernie sat at seat number 6, where did Colin sit?

Nudge I backwards

Rich Richard had a regular hexagonal swimming pool at the back of his multi-million dollar mansion. A triangle of planks had been placed across the pool by joining every second corner of the hexagon. If the hexagon has side of length $\sqrt{12}$ metres, what is the length of each plank (metres)?

Nudge I backwards

Arrogant Arnold was the top mathematician in his school and let everyone know. 'Nothing is an interesting number. Whatever you multiply it by you always get nothing.' 'What about with fractions?' asked Wiseman. 'Even with fractions' confirmed Arnold. Wiseman was not convinced. 'I can think of an exception. How about three-quarters of none?' What answer did Wiseman have in mind?

Nudge 2 forwards

In excavations in the Outer Moogles, engraved tablets have been discovered with arithmetic calculations. Unfortunately, in the one below, some of the characters have been eroded. Assuming that the ? and ! are different signs and the two missing digits are identical, what digit x is missing?

$$10(10 ? x) = 35 ! x$$

Solutions p.91

The stuff of dreams

One night, as Tiny Tum was tucked up in bed, he dreamt he was in Noddyland. It was a land of chocolate clouds and ice cream meadows – anything was possible! Tiny Tum found himself sitting by a milkshake stream. It would be great to have a pint of milkshake but he quickly realised that there was nothing to carry it in.

Suddenly, two cylindrical beakers appeared. One could carry up to three pints and the other up to 10 pints. The former was half the height of the latter, and both were made from impenetrable material having zero thickness! Tiny Tum filled one of the beakers to the brim once only, and using the other beaker, measured out one pint.

What was curious about his method was that, throughout the measuring process, the original quantity remained in the beakers – that is, no milkshake was thrown out!

How was one pint measured?

Counting sheep

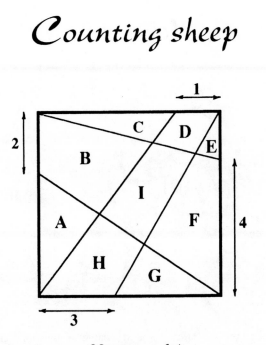

(Not to scale)

Down on Dungdrop Farm, the farmer was trying to count his sheep. However, rather than chase them around the field to count them all, he realised that he could calculate the total number from the plan of his field.

The square field is shown above, partitioned by fences into nine pens. The number of sheep in each pen is proportional to the area of the pen. (A fraction of a sheep can occur in a pen if one sticks its head through the fence!) Apart from the relative measurements shown in the diagram, the farmer knew two other facts: the area of pen I equalled the total area of pens A,C, E,G; and the total number of sheep in pens E,F,G was 49.

How many sheep were in the field?

The reclusive inventor

There was once a reclusive inventor who became so annoyed with unwanted visitors ringing his doorbell that he decided to discourage them by inventing a new one.

The device consisted of a row of six push buttons mounted on the front door, wired in such a way that only one of the buttons would ring the bell. If an incorrect button was pressed (even simultaneously with the correct one) the bell would be temporarily deactivated.

Only his close friends were told the identity of the correct button. Everyone else had to deduce it from an inscription on the front door which read 'Exactly one button is somewhere to the left of the one, that is three to the right of the one, that is somewhere to the right of the one, that is next to the one, that is two away from the one that is first mentioned. Ring the only button of the six that is not mentioned above.'

What was the position of the correct button?

Hints p.82   *Solutions p.100*

Advanced puzzles

The engineer's dilemma

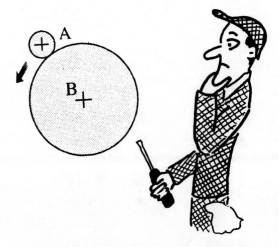

Arnold the Engineer is stumped. In the diagram, the radius of the large wheel is four times the diameter of the small wheel. The centre A of the small wheel rotates anti-clockwise around the centre B of the large one at 16 revolutions per second, so that the small wheel rolls along the surface of the large one. The large wheel also rotates anti-clockwise about its centre B at N revolutions per second.

Now Arnold requires that the circumference of the small wheel does not rotate (that is, the horizontal line through the centre A remains horizontal) as the small wheel moves around the large one.

What value of N must Arnold choose?

Hints p.84   *Solutions p.111*

Lost in space

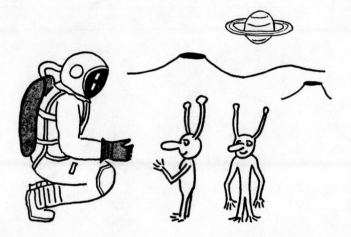

As Captain Klot stepped out of his spacecraft, he was welcomed by five Tiddlybons, the resident space people of Sigma 2. Knowing that the population had a precise hierarchy, Klot asked five questions to try to discover their ranking order. To help identify them, he stuck a label on each one bearing a letter of the alphabet. (The Tiddlybons thought it was a form of greeting and chortled merrily.) Was A higher than C? Was B higher than E? Was C higher than D? Was D higher than B? Was E higher than A? For each question, the second Tiddlybon mentioned whispered a 'Yes' or 'No' to the first mentioned who reported 'Yes' or 'No' to Klot.

Unfortunately, exactly two of the five lied consistently, both in whispering and reporting answers. As it turned out, if all had given truthful answers, their order would have been completely determined. However, precisely two reported answers did not correspond to the true hierarchy. The reported answers were 'No','No','Yes','No','No', respectively. The only Tiddlybon that we can be sure to trust reported a correct answer.

What was their ranking order and who lied?

A tall story

The Android, Bizarre and Clone families all consist of dwarfs and giants. Each of the families has at least one and at most ten of each type, with the total numbers of dwarfs and giants being equal. The giants are much heavier than the dwarfs, each weighing the same whole number squared times the weight of a dwarf. Consequently, the families with the least and the most members have an equal total weight. Now, the Clone family is notoriously mischievous, and one night a third of them each kidnapped one person from the Androids' castle and locked them in the Bizarres' castle. This made the number of occupants in the two family castles equal. When everyone had been returned, the same Clones did it again, but this time each kidnapped one person from the Bizarres' castle and locked them in the Androids' castle. This meant that there were now twice as many occupants in the latter castle as in the former. When the local police heard of this they wanted to know the numbers of dwarfs in each family. What are these numbers?

Hints p.85   *Solutions p.101*

The mathematical garden

Professor Brainbloom had a very curious mathematical garden. There were probability trees, random flowers that had grown from seed, and broken garden gnomes that looked fractionally vulgar! It was an ideal place for thinking, and when the professor had a particularly tricky calculation, he would take a chair outside and use the log table.

Mathematical phenomena used to occur in the garden too. Like the time when the professor was busy laying a path of square slabs. He was just lowering the last slab into position when a number of prime pears struck his head. The slab crashed to the ground and shattered into pieces.

However, far from being annoyed, the professor was delighted. He quickly realised that the square slab had broken into exactly nine triangular pieces so that every angle was less than a right angle.

How were the triangles arranged in the square?

Hints p.84   *Solutions p.116*

Round the clock

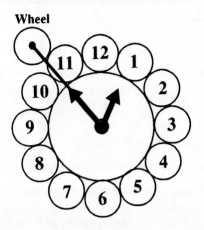

Wheel

Ticktock Town has a clock which long resides in the memory of all who visit there. As shown above, the clock consists of a large circular plate upon which the hour and minute hands rotate. Around the circumference of the plate are 12 small circular plates. These are numbered and arranged so that each touches both of its neighbours and also the large plate.

Now what is extraordinary about this timepiece is that it has a wheel that runs around the outside of the clockface. The wheel has the same radius as each of the small plates, and has an axle attached by a rod and spring to the minute hand. The spring keeps tension on the wheel in the direction of the clock centre so that, as the minute hand rotates, the wheel rolls along the outer contours of the 12 plates without losing surface contact.

How many times does the wheel rotate with respect to the clock centre each hour?

$\mathcal{P}$et hate

House	First name	Second name	Pet	Pet name
2	Herbie	Tipple	Tarantula	Grumpy
4	Judy	Blip	Cobra	Stomper
6	Angus	McTumble	Elephant	Bubbles
8	Celia	Grout	Rhinoceros	Nasher

The children who live in Cobble Street have some unusual pets. Constable L. O. Hullo was so anxious that he made a list of the child's name, pet and pet's name for each house number. However, although he wrote each item in the correct column, he only managed to get exactly one item in each of the four columns correctly positioned.

Back at the police station, the constable tried to correct the list from memory. Now, either the rhinoceros or the tarantula was at number 6; the tarantula was not at number 2. Angus, who had neither a cobra nor a pet called Bubbles or Grumpy, lived at a house number which was two less than McTumble's, who had neither a rhinoceros nor a tarantula. Herbie Blip had a pet called Stomper which was neither an elephant nor a cobra.

Can you give the correct child's names, pet and pet's name for each house number?

Hints p.83   *Solutions p.106*

Sum secret

Oliver Oddwelly was a cringing wreck of a man. He frequently needed to have his safe combination number, his bank account number, and his credit card number on hand, but was too frightened to write them down anywhere in case someone found them. However, one day, just as he was oiling the padlock on the refrigerator, he suddenly hit upon an ingenious way of concealing the numbers. He decided to compose an arithmetic problem, the solution to which would reveal the three seven-digit numbers he needed to remember.

The sum he invented is shown. One digit can be erased from each row (not necessarily the same position in each row) and the gaps closed up to leave three columns of digits, then a second digit can be rubbed out in the same way to leave two columns, then a third to leave one column, so that a valid sum remains each time. The three sets of seven digits erased (read down the columns) respectively reveal the numbers he had to remember.

What were the three numbers?

Hints p.84   *Solutions p.111*

Selfish sons

A farmer had fallen on hard times and in desperation he called his sons in from the fields to discuss what money each could give to buy food for the family. However, he had a problem for he knew that anyone who had money would lie, not only about his own wealth but also about what the other sons possessed and had said.

The first son reported to his father that the third had said 'Precisely one of my four brothers has money'. The second claimed that the fifth had said 'Exactly two of my four brothers have money'. The third son told his father that the fourth son had said 'Precisely three of my four brothers have money'. The fourth reported that the second had told him 'All of my four brothers have money'. The fifth son reported that the third had money and also told his father that the first had admitted having some. Luckily, the farmer was no fool and managed to work out who was lying.

Which sons had money?

Hints p.85   *Solutions p.113*

The Three Bears

The Three Bears were sitting down to eat their porridge. It should have been a happy family occasion, but it soon became apparent to everyone that something was wrong. They looked at each other in silence, drew a deep breath, then in earth-trembling chorus growled 'Who's been eating my porridge?!'

It was a case that would have delighted a psychoanalyst, for the bears had eaten each other's porridge!

Now Baby Bear originally had 30 spoonfuls of porridge in his bowl, Mummy Bear had 60 and Daddy Bear had 90.

Baby Bear had eaten half the amount that Mummy Bear would have had left in her bowl if only Daddy Bear had eaten from her bowl and had eaten one third of his actual total consumption.

Mummy Bear had eaten half the amount that Daddy Bear would have had left in his bowl if only he had eaten from his bowl and no-one else.

Daddy Bear had eaten the amount that Baby Bear would have had left if only Mummy Bear had eaten from Baby Bear's bowl and had eaten one third of her actual total consumption.

How many spoonfuls of porridge had each bear eaten?

Hints p.85   *Solutions p.116*

Cryptic cave lines

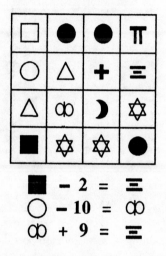

While exploring a cave one afternoon, Hogman Hunter uncovered a strange drawing on the ground. Brushing aside the dust, he also found three equations showing relationships between some of the symbols in the drawing. After thinking for a while, Hogman came to the conclusion that each symbol in the grid represented a letter in the alphabet, so that no two different symbols had the same letter.

When the correct letters had been substituted for the symbols, one could expect to see eight four-letter words, four reading from top to bottom, and four from left to right. Evidently, the equations gave the relative positions of letters in the alphabet. Suddenly, disaster struck. The roof fell in by the cave entrance blocking Hunter's exit. 'Calamitous!' exclaimed Hunter, but he could just as appositely have said one of the words in the grid.

What are the words in the four rows?

The three prisoners

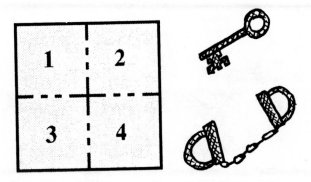

A prison block had four cells arranged in a square as shown. Barred windows between adjacent cells allowed each of the three prisoners to see into his two neighbouring cells.

The blind warder was forgetful and could not remember which cells were occupied. So, starting at cell 1, he visited each cell, in no particular order, to convey the demand that, on his signal, everyone in a cell must make at least one statement about his perceived state of occupancy of an adjacent cell.

On his fourth visit, he gave the signal, and four statements echoed around the block:

(a) 'I can see a prisoner in cell 2'
(b) 'I cannot see a prisoner in cell 1'
(c) 'I cannot see a prisoner in cell 2'
(d) 'I cannot see a prisoner in cell 4'

When reporting on an adjacent cell, exactly one prisoner would always lie about his perception, and no two people were in the same cell when the statements were made.

Which cell had no prisoner, and can you identify the false statement(s)?

Taking a bath

In Texas at the turn of the century there was a wash house which had four identical large baths. These the owner would fill according to the wishes of the customer. One afternoon the four Badboy brothers showed up demanding a bath each, to be prepared in such a way that when each man totally submerged himself (as was their custom on climbing in) the water would just reach the brim.

Knowing that Dan was twice Clancy's volume but half Ben's, and Abe was one and a half times Dan's volume, the owner poured a different quantity of water into each man's bath. Now the Badboys were feeling mighty mean, so they all climbed into an incorrect bath, one to a bath, causing at least one to overflow. They then accused the owner of incompetence and threatened that unless the water just reached the brim in all of the baths when they climbed into their correct ones, he would have something really nasty done to him.

The owner examined the baths and realised that two of them each contained water with a volume less than that of the smallest man. Furthermore, the total water left in those baths that needed topping up was equal to the total additional water required.

Who had climbed into whose bath?

The balanced bridge

Four men wish to cross a river, two from the south side and two from the north side. To do this they have to use a rotating balance bridge.

The bridge works thus: an equal weight of men stand in a carrier on each end of the bridge and then a lever is pulled on the south side to rotate the bridge, in a horizontal plane, one half-turn about a point midway along its length. This allows both sets of men on the bridge to travel to the other side.

However, the four men are faced with a dilemma. They realise that there is no arrangement of the men on the bridge that will permit them all to cross since there always has to be someone on the south bank to operate the lever.

Suddenly, they see a man walking his dog on the north bank. They enlist his help, and in two separate half-turns of the bridge the two pairs of men reach the other side. The extra man then continues his walk on the north bank.

The five men involved weigh 1,2,3,4 and 5 units. The first lever operator was lighter than the second one, and the sum of their weights equalled that of the dog owner.

How were the men arranged on the bridge on the north and south sides for the two journeys?

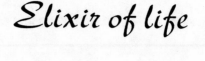

Elixir of life

There is said to be a valley where the inhabitants once had the elixir of life. Long ago, in the Neo-Hermet period, when man dabbled in alchemy, there lived a civilisation who could dramatically increase their life expectancy.

An ancient Greek scribe called Menpet visited the tribe and asked an old man how old he was. The old man wanted to protect their secret and said that his age could be found by taking the product of two numbers: the sum of the three digits in his age, and the product of the same three different digits. The scribe asked if there was a quicker way of finding it and the man rather cryptically replied 'The digits can be found by deleting 'em.'

How old was the man?

$\mathcal{T}$he great escape

A prisoner sits in his cell planning an escape. The guard arrives, checks that the prisoner is still there, and then leaves. The prisoner now has to carry out his escape plan for he knows that when the guard returns he will be executed.

The cell is situated at the beginning of a long straight corridor which is partitioned by five doors. The doors operate on different time switches so that the first, which separates the cell from the corridor, opens every 1 minute 45 seconds, the second every 1 minute 10 seconds, the third every 2 minutes 55 seconds, the fourth every 2 minutes 20 seconds, and the fifth, which is at the end of the corridor, every 35 seconds. Every once in a while, the five doors open simultaneously. When this happens, the guard arrives, looks down the corridor to check the cell, and then leaves.

The prisoner has calculated that in making his escape it will take 20 seconds to cover the distance between consecutive doors, which is longer than the amount of time a door stays open. He also knows that if he stays in the corridor for longer than two and a half minutes an alarm will sound so the prisoner plans to escape in the shortest possible time.

How long before the guard returns does the prisoner clear the last door?

Logical legacy

Tom Bodger was an eccentric to the last, and when he died he left precise instructions with his lawyer regarding the amount of money his only son should receive. The old man had devised a test for his son which would determine his inheritance.

The lawyer presented the son with six coloured boxes: two blue, two green, and two red, and was told that each box contained a sum of money. Two of the boxes contained $10 000 each, two contained $15 000 each, and two contained $25 000 each. He was allowed to choose any two boxes of the same colour, the total contents of which would constitute his entitlement.

To help him decide, each box had a statement engraved on it. The blue boxes stated that 'Both a blue box and a red box contain $10 000 each'; the green boxes stated that 'Both a green box and a blue box contain $25 000 each'; and the red boxes stated that 'Both a red box and a green box contain $15 000 each'.

Only one of the three statements was true and the corresponding two engraved boxes contained the greatest total of the three possible pairs.

What was the total contents of each pair?

Hints p.85   *Solutions p.117*

Wire wizards

Roadworkers found eight wire ends protruding from a pipe in London. In Glasgow they discovered the other ends of the eight wires. Two foremen, Smith and Campbell, met to discuss how to match up the two sets of ends.

Back in London, Smith took a battery and connected pre-arranged numbers of ends to the positive terminal, the negative terminal, and left at least one wire free. In Glasgow, Campbell labelled his ends A to H, then with a bulb tested each pair of wires that could be formed from the eight, for a circuit. Knowing the pre-arranged numbers Campbell could identify wires in each group. The idea now was for Smith to disconnect the battery, and Campbell to join six of his ends into three pairs, then tell Smith which ends he'd joined and which wires were in each group. Smith could then test all pairs of his ends using his battery and bulb, and thereby correctly identify his wire ends.

But when Smith phoned Glasgow, Campbell could only recall that F made at least one circuit with another wire but not with G; B made a circuit with H; there were at least 6 pairs involving A which did not make a circuit and the connections made were A to G, B to C, and E to F. Smith tested all pairs of wires, and correctly labelled the ends.

What were the wires in each group?

Hints p.85   *Solutions p.114*

The broken pentomino

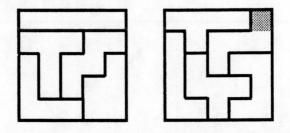

Toddler Toby had been playing on the floor with his 12 pentominoes (a set of five squares joined along their edges). What joy! After much effort, the youngster had succeeded in making two 5 × 5 squares from 10 of them.

Delighted by his discovery, he ran to tell Mummy. Unfortunately, in her haste, Mummy trod on them, sending them scattering and breaking a square off the Z-shaped piece (shaded square in diagram).

'Not to worry. Mummy will fix it!' She certainly fixed it, but glued the square back in the wrong place to make an entirely different pentomino to the 10 used. Luckily, it did not matter, for Toddler Toby still managed to make two 5 × 5 squares from the material used, with exactly five pentominoes positioned as before.

How is it possible?

Hints p.84   *Solutions p.108*

One for the road

Five drunks were trying to push their car home. Being the worse for drink, they often found themselves pushing against one another, with some pushing at the front of the car and some at the back. On one occasion, the twins who each had the least strength managed to move the car with only one of the other twins pushing against them. On another, the stronger twins, each with the same strength, moved the car despite being opposed by only the strongest man, who never pushed at the front.

Being too drunk, one from each of the twins gave up and began to walk home. The three remaining men now had 10 ways of pushing the car forwards by arranging themselves at the front and the back, if no more than one man rested with any arrangement.

Eventually, they chose the most effective arrangement where a man pushed at the front. The strength of each of the men was a whole number and the arrangement they chose had a resultant strength of 11.

What were the strengths of the three men?

Hints p.83　　　　　　　　　　　　*Solutions p.108*

In the same boat

Twelve special agents, all chosen to have an equal weight, were sitting in a small boat ready to embark on a secret mission. In order to balance the boat in the water ABCD sat at the back of the boat, EFGH in the middle, and IJKL at the front. However, before leaving, their group leader looked over the side and noticed that the boat tilted forwards. Assuming that someone in the boat must be an enemy agent with a different weight from the rest, he consulted his manual and found that for such a situation two tests could be carried out. These would not only identify the imposter but would also show whether he was heavier or lighter than the chosen weight. Each test consisted of a rearrangement of the 12, keeping four in each group, and noticing the resulting inclination of the boat.

For the first test only, four interchanges were made involving ABCEFGIL so that A and G then occupied the same group. Noting the inclination of the boat, the leader rearranged the crew for the second test, without then changing his own position, so that ACHL finished at the back of the boat, DIJK in the middle, and BEFG at the front. From the resulting inclination he managed to deduce who the enemy agent was, and to his great surprise found that it was himself! Mistrusting his own logic he ordered the mission to go ahead and two of the crew were subsequently caught trying to send a radio message to the enemy. One was heavier than the chosen weight by as much as the other was lighter.

Who were the two enemy agents?

Hints p.84 *Solutions p.102*

Hints

Popular puzzles

Word bandit 3
Carefree.

True to the tribe
Assume that Cowa tells lies. Then three possibilities arise.

(1)   Appu is truthful and Dobi lies.
(2)   Appu lies and Dobi is truthful.
(3)   Appu and Dobi are both liars.

If none of these work then assume Cowa tells the truth.

Muddle market
Either Dippy buys from the fishmonger and Grumble the greengrocer, or Dippy from the butcher and Grumble from the confectioner.

The horse and the hurdle
The horse actually contains part of a hurdle and vice-versa.

Railway rhyme
What line(s)?

Cornflakes and porridge
Label the inmates A to G and write out the relations as pieces of a jigsaw puzzle, e.g. if Slasher Sam is A and Desperate Deborah is B then we have B-A.

Strange street
The couple can only live next door to the hippy in two ways. Try out each way and see what deductions result.

Crumple towers

The flat below Nosebag has neither Cesspot nor Bucket, then it must be Hump. Bertha is not Nosebag, who is male, and is not Hump who has 4 pets.

The three piles of coins

Take one of the coins and propel it along the table.

Quite a card

Since a green card must occupy one of the corners in the third column then the corners in the first column must be yellow.

Find the burglars

If Peterson burgled Catalan (condition 5) then Bloggs burgled Belvedere (4). Smith/Harper did Oakmoor/Pinewood (6) so Jones did Ardswick.

The doubtful die

Try drawing out each die as a flat map to be folded into a cube.

Cubic incapacity

To find the delivered quantity, we want one of the whole numbers from 1 to 10 so that when cubed and subtracted from 3025 produces the square of some whole number (greater than 10).

Fractionally mean

One way to tackle the problem is to work out what the last two digits are and work from the end to the beginning.

Signs of confusion

If Booliba to Rumba is 4 then Rumba to Booliba cannot be 2. One or both of these numbers must be wrong.

The enigmatic interview
There are two face-up reds so red can be eliminated from the statements.

Extra sensory deception
Treat the relations as pieces of a jigsaw, draw them out and see how they can be pieced together.

Wong sum
Try crossing out each digit on the left in turn then seeing if the left-hand side equals three of the four digits on the right.

The six sheep pens
No fence belongs exclusively to one pen.

Domino chain
Altogether there are 28 tiles totalling 168.

Room for assistance
Only Splutter, Britches and Crumpet could be in the morning room. Which assumption leads to a complete deduction?

Safe conduct
There are only three combinations for the second, third and fourth digits.

Prime of life
Let the digits in grandma's age be A and B so that her age is $10A + B$.

Save the city
This can be tackled systematically by working from right to left on the bottom row, covering each digit in turn to see which two digits can be deleted in the rows above.

Word bandit 5
Weekly.

Harvey's wall climber
Remember that cog-wheels in contact rotate in opposite directions.

Word bandit 2
Gold coin.

Word bandit 1
Make it known.

Word bandit 4
Inopportunely said.

Romance on the stone
This is a matter of trying out both directions.

The reclusive inventor
Only five buttons are mentioned and the first mentioned and last mentioned are the same one. Try starting with the 'three to the right of' relation.

The stuff of dreams
Half of the volume of the large beaker is five pints and the volume of the small one is three pints. How can filling the small beaker be used with this fact?

King-size conundrum
Think in areas.

Sorry Dad
What is being taken home and from what word?

Counting sheep

Let the length of the field be L and work out the areas of triangles.

Advanced puzzles

Lost in space

With the particular questions asked, a completely determined order only arises from having exactly one 'Yes' or one 'No' in five truthful answers.

Taking a bath

Let Clancy have volume V. Then Dan, Abe and Ben have volumes $2V$, $3V$ and $4V$, respectively. Let Ben have a volume V_0 of water in his bath initially. Then each bath has capacity $V_0 + 4V$, and Abe, Dan and Clancy have $V_0 + V$, $V_0 + 2V$ and $V_0 + 3V$ in their baths, respectively.

Elixir of life

There are only two names in the puzzle so why not try deleting 'em!

Round the clock

There are two components to the wheel rotation: rolling along a path equal in length to the curved path and rotation of the tangent to the curved path at the point of contact.

Pet hate

Angus will not fit at number 8 and cannot be at 4 because then number 6 has no possible pet.

One for the road

There are 10 ways of arranging the men. Try placing their strengths in order.

The great escape

The doors have a cycle which is a multiple of 35 seconds.

The balanced bridge

Since the dog owner starts and finishes on the same bank, the two pairs of men, having exchanged sides using a balance, must both have an equal weight. Hence the dog owner has weight 1, 3 or 5.

The mathematical garden

Each corner of the square must be divided by at least one line and every meeting point for lines inside the square must have at least five lines. Where can you fit a pentagon?

The engineer's dilemma

There are two components to consider: the rotation of the circumference of A due to the rotation of B, and its rotation due to its motion around B.

The broken pentomino

The new pentomino must be flipped over to get a solution.

Sum secret

Try separating the sum into three separate sums: lines 1–3, 3–5 and 5–7. Analyse each one and look for a joining thread.

Cryptic cave lines

Hunter's exclamation could be giving you a ride, anagrammatically speaking.

In the same boat

How can the two tests cooperate to identify the group leader? Match each possible inclination to a group of suspects.

Selfish sons

Whoever has money lies about his own wealth so the first son would not admit having money whether he had some or not.

The three prisoners

Who else is in a cell when the warder gives the signal?

The Three Bears

Let the amounts that these bears originally had in their bowls be D, M and B respectively, and their total consumption be d, m and b respectively. Now try to form the three equations.

Wire wizards

If Glasgow has X wires at one terminal, Y at the other, and Z free, the circuit tests in Glasgow must show that X wires make a circuit with Y others, Y wires make a circuit with X others, Z wires make a circuit with no others. So X and Y must be different in order to distinguish between the groups. Now consider how the identity of the groups can be used to identify connections.

Logical legacy

In turn, assume each statement to be true and the others false. Can the corresponding colour contain the highest total of the three?

A tall story

Two linear equations involving A, C and B, C can be formed, where A,B,C are the total Androids, Bizarres and Clones, respectively.

Solutions

Popular puzzles

Safe conduct

The correct combination is 65292. Since the third digit is three less than the second, and the fourth is four greater than the second, there are only three possible combinations for the second, third and fourth digits. These are –307–, –418–, and –529–. With the first digit three times the fifth, the only possible combinations for the first and fifth digits are 0 – – – 0, 3 – – – 1, 6 – – – 2, and 9 – – – 3. The solution arises from combining these two sets of possibilities, with the added criteria that there are three combinations of two digits that that each sum to 11.

Word bandit 1

The 10-letter word (with letters in reverse order) is ETAGLUMORP.

(1) His younger brother was 6. Let his age be x, Dimple's age be y, and his father's age be z. Then $z=3y$, $z=x^2$, $y=2x$. So $6x=x^2$ and $x=6$. (At 6, E to G.)

(2) The polygon has 8 sides. (At 8, U to T.)

(3) The number is 4. FOUR is the only digit to have as many letters as its value. (At 4, U stays U.)

(4) The missing digit is 2. The equation reads
$$62/2 = 29 + 2$$
(At 2, P to O.)

(5) The factor is 5. The number $5400=2 \times 2 \times 2 \times 3 \times 3 \times 3 \times 5 \times 5$ so that a factor of 5 is required to make the next nearest cube. (At 5, M to L.)

Fractionally mean

Grabber rubbed out pairs (top,bottom) of digits in the order: (4,2), (6,3), (7,8) to leave 167/835, 17/85, 1/5. This is the only Sequential Digital Deletion Fraction (SDDF) of rank (4,4) – i.e. four digits in the numerator and denominator, respectively, with deletion pairs occurring in sequence – where all the digits are different.

Domoryad (1964) has given examples of non-sequential Digital Deletion Fractions where the same group of digits are deleted simultaneously in the numerator and denominator. For example, in the following DDF of rank (6,8) we delete 18 so that

$$\frac{143185}{17018560} = \frac{1435}{170560}$$

He also gives an example of a SDDF of rank (4,4), however, the reduction sequence is not unique:

$$\frac{2666}{6665} = \frac{266}{665} = \frac{26}{65} = \frac{2}{5}$$

Here, all deleted digits are identical and there are several possible deletion sequences. In contrast, the SDDF in the present puzzle relies on there being only one deletion sequence.

King-size conundrum

The path was 5000 paces long. If the path is straightened out into a line, its area equals the area of the whole square, that is, 100×100 square paces. If this is divided by the width of the path, its length results.

The enigmatic interview

Globule has yellow, Crust has yellow, Wilson-Bucket has blue, Nostril has green.

For Globule, the possible colours are green or blue for a true statement, and yellow for a false one; Crust has yellow for true, and blue for false; Wilson-Bucket has blue or yellow for true, and green for false; and Nostril has yellow for true and green for false.

From these we aim for a green, blue, yellow, yellow combination from two true and two false statements. Suppose Wilson-Bucket lies. Then he has green, Nostril must have yellow and is truthful, leaving two yellows or two blues for the other two cards with one truthful and the other a liar. This is not possible since we must have a yellow and a blue. Hence Wilson-Bucket is truthful with blue or yellow.

Suppose Wilson-Bucket has yellow. Then Nostril must lie with green and the remaining yellow and blue cannot fit in with the remaining liar and truth-teller. So Wilson-Bucket has blue. Then Crust must tell the truth with yellow and we have Globule lying with yellow and Nostril lying with green.

The three piles of coins

Pick up the top coin of pile 5, place it on the table on the right of and in line with the three piles, and give it a strong shove towards the remainder of pile 5. The bottom coin of pile 5 will be knocked out leaving the impinging coin at 6 and the original middle coin of pile 5 at 5. The knocked out coin removes the bottom coin of pile 3 and comes to rest at 4. The top coin of pile 3 falls to 3 and the new knocked out coin comes to rest at 2.

Word bandit 5

The 10-letter word (with letters in reverse order) is LADAMODBEH.

(1) Nine times. Stretch out the circumference of the large coin into a straight line. This line is eight times the circumference of the small coin (noting the difference between radius and diameter) so the small coin revolves eight times as it runs along it. When the straight length is wrapped around into a circle, this path that the small coin runs on rotates 360 degrees, so one revolution must be added. (At 9, N to L.)

(2) Colin is at 3. If Andrew sat two to the left of Deborah and Barbara was somewhere to the right of Deborah we have A–DB–, A–D–B or –A–DB. If Ernie was not next to Deborah this gives ACDBE or EACDB. With Ernie at 6, in the second case Barbara would be at 10 which is illegal. So we have ACDBE with Colin at 3. (At 3, E to D.)

(3) Each plank has length 6 metres. Note that $\sqrt{12} = 2\sqrt{3}$. Join the centre of the hexagon to each corner. By Pythagoras's theorem, half of the length of the plank is $(\sqrt{3}/2)(2\sqrt{3})$. (At 6, B to A.)

(4) Wiseman was thinking of one. Its three letters make up three-quarters of the four-letter word 'none'. (At 1, F to E.)

(5) The missing digit is 4. The signs are $+$ and $\times$. (At 4, M to O.)

The doubtful die

Die B is the odd one out. Watch out for the orientation of the three spots.

Strange street

The woman is at 5, the skinhead at 7, the hippy at 9, and the couple at 11.

Cornflakes and porridge

Cornflake Colin's neighbours are Vicious Vince and Poisonous Pat. The order of the prisoners from 1 to 7 is Vicious Vince, Cornflake Colin, Poisonous Pat, Harry the Hatchet, Deperate Deborah, Gruesome Gertie and Slasher Sam.

Let us arbitrarily label these inmates A to G, respectively. Then we have either CD or DC, either EFG or E–GF, either D– –A or A– –D, and either E–C or C–E. Combining the second and fourth possibilities with the fact that none of the seven cells are shared, we have either C–EFG or C–E–GF. The first possibilities give CDEFG, CDE–GF, DC–EFG, or DC–E–GF. The third possibilities then limit the options to only one solution A–CDEFG.

Signs of confusion

Wobble has inhabitants who always tell the truth, at Booliba they only tell the truth half of the time, and Rumba is full of liars. Take each sign in turn and assume that it has two correct numbers. Then only one of the four numbers in the other two signs can be correct. Only the Wobble sign satisfies this condition.

Find the burglars

Jones stole trophies from Ardswick, Smith took money from Pinewood, Peterson pilfered paintings from Catalan, Bloggs nicked antiques from Belvedere, and Harper had jewellery from Oakmoor.

If Peterson burgled Catalan (condition 5) then Bloggs burgled Belvedere (4). Smith/Harper did Oakmoor/Pinewood (6) so Jones did Ardswick. Jones steals trophies (3) so Peterson has paintings (1) and Smith steals money (2). Smith cannot therefore be at Oakmoor which has jewellery (7) so is at

Pinewood. Harper must be at Oakmoor with jewellery and the antiques with Bloggs at Belvedere.

Word bandit 4

The 10-letter word (with letters in reverse order) is SOPORPALAM.

(1) Eight triangles. (At 8, O stays O.)
(2) The missing digit is 3 with a minus and division sign. Note that 15 is also a solution but is not a single digit. (At 3, B to A.)
(3) The digits total 5. Let the two digits be a and b. Then the spots must be $a+b$, $7-a$, $7-b$ so that their sum is 14. (At 5, R stays R.)
(4) Four ways. (At 4, O to P.)
(5) Zero distance. The right-angled triangle has two 45 degree angles so the distances from Gunslinger to Smokeshot and Smokeshot to Deadgun are equal. (At 0, L to M.)

Quite a card

The layout of the rows is yellow, red, green; blue, blue, green; and yellow, green, yellow. The corners in the first column must be yellow since a green card occupies one of the corners in the third column. Both of the corners in the third column cannot have a green card since there would be no place for the third yellow. So since the third column has exactly two green cards, a green card must be in the second row, third column. This leaves the other two positions in the second row occupied by blue cards. Of the two corners in the third column, one is yellow, the other green. Hence the red card must be in the first row, second column. The first row green card can only be in the third column, leaving a yellow card in the third row, third column. So the final green card can only be in the third row, second column.

True to the tribe

Only Appu, Cowa, and Dobi are honest. Assume that Cowa tells lies. Then three possibilities arise.

(1) Appu is truthful and Dobi lies.

(2)   Appu lies and Dobi is truthful.
(3)   Appu and Dobi are both liars.

None of these possibilities produce exactly two liars. So Cowa must be truthful. This means that both Appu and Dobi are also truthful, and that Babble and Eva are both liars.

Harvey's wall climber
Wheels A and B both turn anti-clockwise. So Harvey wasn't so screwy after all!

The horse and the hurdle
Note the change in the line of sight.

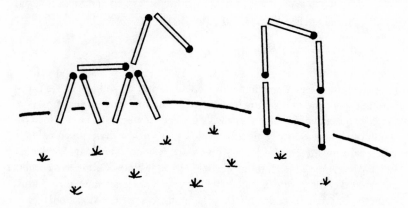

Domino chain
The remaining dominoes are [2,4] [4,0] and [0,3]. There must be 28 combination pairs in a domino set with digits totalling 168. This means that three tiles remain totalling 13. Two of the six digits must be 2 and 3 to match the chain ends. This leaves a total of 8 for the other four squares. Since domino ends must match, the illustration shows the possibilities. The fact that no two dominoes can be alike gives the solution.

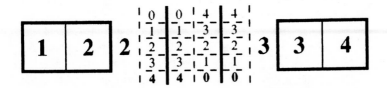

Sorry dad

His name was Tom. Remove 'her' from 'Mother' then look at what remains from the back.

Romance on the stone

Frederick should keep jumping clockwise so that the rendezvous takes place in nine jumps each. When Frederick jumps anti-clockwise it takes ten jumps each.

Counting sheep

There were 245 sheep in the field. Let the length of the field be L. Then the areas of pens A + B + C, C + D + E, E + F + G, A + H + G are respectively $L(L - 1)/2$, $L(L - 4)/2$, $L(L - 3)/2$, $L(L - 2)/2$. The total of these four areas is $L(2L - 5)$. If we now introduce the given condition that area A + C + E + G = I, then the total given above becomes that of the whole field, that is, L^2. This means that $L = 5$ and the area of the whole field is 25. So the area of E + F + G is 5 and we have the proportionality

$$5/25 = 49/N$$

where N is the number required. This leaves us with $N=245$.

Word bandit 2

The 10-letter word with (with letters in reverse order) is DNARREGURK.

(1) Six ways. (At 6, S to R.)
(2) Nine times. The area of the triangle with the hypoteneuse equal to 1 is 1/4 cm^2. So $128(1/2)^n = 1/4$, so $n=9$. (At 9, E to D.)

(3) Four o'clock. The number of hours between the two noons is $3x$, where x is the number of hours to noon today. Since this equals 24 hours, $x = 8$ and the hour is 4 a.m. (At 4, C to E.)

(4) The digit is 8. The digits that appear in neither their square nor cube are 2,3,7,8, while the digits that appear in both are 0,1,5,6. The only two digits that satisfy the condition are 8 and 5. (At 8, M to N.)

(5) Two thousand dollars. Let the money in the safe be x and that in the cash register y. Then $x - y = 10$ and $(x + y)/2 = 7$. So $x = 12$ and $y = 2$. (At 2, S to U.)

Wong sum

The three equations were as follows:
$$3(43 + 02) = 135$$
$$3(3 + 02) = 15$$
$$(3 + 02) = 5$$
Note that 02 is a perfectly valid way of writing 2.

Prime of life

Grandma was 72 years old. Let the two digits be A and B. Then $10A + B - 10B - A = 45$, so $9(A - B) = 45$. This gives $A - B = 5$ which is only satisfied by the pairs (5,0), (6,1), (7,2), (8,3), (9,4). Only the third pair are both prime.

Room for assistance

Assuming that Splutter was in the morning room gives Twinkle in the library, Uppity in the conservatory, Lipstick in the study, Britches in the dining room, and Crumpet in the kitchen. The other two possibilities for the morning room are Britches and Crumpet. Assuming the first gives Crumpet in the kitchen but yields no further information. Assuming the second gives no deductions whatsoever.

The six sheep pens

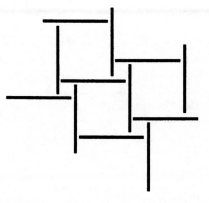

Crumple Towers

Floor 4 had Agatha Cesspot who paid on a Monday with 2 pets; floor 3 had Cuthbert Nosebag who paid on a Friday with 1 pet; floor 2 had Sidney Hump who paid on a Wednesday with 4 pets; and floor 1 had Bertha Bucket who paid on Tuesday with 3 pets.

Since the flat below Nosebag has neither Cesspot nor Bucket, then it must be Hump. Bertha is not Nosebag, who is male, and is not Hump who has 4 pets. Bertha cannot be Cesspot who paid her rent on Monday or Wednesday, so Agatha is Miss Cesspot, and Bertha is Bucket. The flat below Nosebag had 4 pets so Cesspot has 2 pets, which leaves Bertha Bucket with 3 and Nosebag with 1. There are only six ways the surnames can be arranged, only three of which give exactly one correct in the list. Only one of those gives exactly one number of pets correct in the list. The first names are easily deducible and the rent days follow from the realisation that either Wednesday or Friday is correct.

Cubic incapacity

There were 216 unit cubes missing.

If we cube each whole number from 1 to 10 and add the results, we get the same answer as adding all the whole numbers from 1 to 10 and squaring the result. The latter calculation gives 55 squared which is 3025. This is the ordered quantity.

To find the delivered quantity, we want one of the whole numbers from 1 to 10 so that when cubed and subtracted from 3025 produces the square of some whole number (greater than 10). There are two candidates: 6 which gives a delivered quantity of 2809, and 10 which gives 2025.

Now, Father and Mother have the two largest whole numbers (9 and 10), and Puddle, who is one of the twins, cannot be either of them. So the unbuilt cube has side 6 so that 6-cubed unit cubes were missing.

The stuff of dreams

Half of the volume of the large beaker is five pints. When the small beaker is seated on the bottom of the large one, the volume around the small beaker up to its brim is five minus three (i.e. two) pints. So the small beaker is filled from the stream and the three pints are poured into the large beaker. The empty small beaker is then placed upright inside the large one, into the milk-shake. The level rises and flows into the small beaker. When it is seated on the bottom of the large beaker, two pints occupy the volume around the small beaker and one pint is inside it. The small beaker with the pint inside is then removed from the larger one.

Railway rhyme

The first letter from each line spells PARIS.

Muddle market

Mrs Folly bought from the greengrocer, Miss Dippy the butcher, Mrs Grumble the confectioner, Mrs Vacant the fishmonger, and Miss Witless the ironmonger. Either Dippy buys from the fishmonger and Grumble the green-grocer, or Dippy from the butcher and Grumble from the confectioner.

In the first case, we eliminate ironmonger and confectioner from Grumble, and butcher and confectioner from Dippy. Also, the fishmonger can now be with neither Folly nor Vacant, and the greengrocer cannot be with Folly. The butcher can only be with Witless but there is nothing to decide how to place Folly and Vacant with the ironmonger and confectioner.

In the second case, Dippy is not with the fishmonger nor confectioner; Grumble is not with the ironmonger nor greengrocer; the butcher is not with Witless; and the confectioner is not with Folly, Dippy, Vacant, nor Witless. This leaves Folly with the greengrocer, eliminating her from the ironmonger and fishmonger, so that Vacant is with the fishmonger and Witless with the ironmonger.

Save the city

The sums generated by the deletions are as follows:

```
  3 2 6        2 6        2
+ 2 6 8      + 2 8      + 2
-------      -----      ---
  5 9 4        5 4        4
```

Further examples of Digital Deletion sums appear elsewhere in Clarke (1994).

Word bandit 3

The 10-letter word (with letters in reverse order) is TNAICUOSNI.

(1) One revolution per second. We find that $50(2.5)=125y$ where y is the number of revolutions per second of the back wheel. So $y=1$. (At 1, L to N.)

(2) Six ways. (At 6, G to I.)

(3) Emily was nine years old. We have the equation $10x + y = 2xy$ from which we get $x = y/(2y - 10)$. For x to be positive, y is 6,7,8 or 9. Only $y=6$ gives a digit, $x=3$ so Herbert is 36 and Emily is 9. (At 9, S to T.)

(4) The garden diameter was 4 cm. Using a point in the centre, the hexagon can be divided into exactly six equilateral triangles. If the length of a triangle's side is x then its height is $(x/2)\sqrt{3}$. This gives the area of the lawn as $6(1/2)x(x/2)\sqrt{3} = \sqrt{108}$. From this x^2

(5)   Rummage took 5 plums. The total number removed has two
       digits, is less than 15, and must be divisible by three in order to
       produce an equal distribution. Only 12 plums removed is
       possible. The only possible numbers removed are 9,2,1; 8,3,1;
       7,4,1; 7,3,2; 6,5,1; 6,4,2; and 5,4,3. Each person must be left with
       4, so Grabbit and Rummage must have at least 4 in order to give
       any to Tumble. Only 6,5,1 satisfies this condition. (At 5, E to C.)

Extra sensory deception

Numbering the squares from 1 to 9 from left to right, in numerical order we
have red, orange, violet, blue, white, turquoise, green, pink, yellow. Since
nine colours are mentioned, each colour is used exactly once. The
red/blue/white combination can be in 1,4,5; 2,5,6; 4,7,8; or 5,8,9. For the
second and third cases, both of the green/yellow and the orange/pink
combinations cannot fit. For the fourth case, the green/yellow can occupy
4,6 with the orange/pink at 1,7 but the turquoise/violet cannot fit. Only the
first case works.

The reclusive inventor

The correct button is fifth from the left.

The inscription states that only one of the six buttons is not mentioned, so
exacty five different buttons are mentioned. In the statement of relations
between positions, the inscription refers to six buttons; however, the first and
last mentioned buttons are the same one.

One now labels the buttons mentioned in the inscription in the order A B
C D E A and begins by noting the most precise relation, that is, B is three to
the right of C. Since no position may have two letters, our deductions show
that from left to right the buttons are D E C A ? B. So the correct button must
be the missing one.

Advanced puzzles

A tall story

The Android family has 10 dwarfs, 4 giants; the Bizarre family has 4 dwarfs, 6 giants; and the Clone family has 1 dwarf, 5 giants.

Let D_a be the number of dwarfs in the Android family, D_b the number in the Bizarre family, D_c the number in the Clone family, G_a be the number of giants in the Android family ..., all positive integers. Then we have

$$0 < D_a, D_b, D_c, G_a, G_b, G_c \leqslant 10 \qquad (1)$$
$$D_a + D_b + D_c = G_a + G_b + G_c \qquad (2)$$

Let

$$D_a + G_a = A$$
$$D_b + G_b = B \qquad (3)$$
$$D_c + G_c = C$$

Then

$$A - C/3 = B + C/3 \qquad (4)$$
$$A + C/3 = 2(B - C/3) \qquad (5)$$

Yields the solutions

$$A = 7C/3 \qquad (6)$$
$$B = 5C/3 \qquad (7)$$

Now (1) and (3) show that

$$1 < A, B, C \leqslant 20$$

are integers so (6) requires that

$$C = 3 \text{ or } 6 \qquad (8)$$

Let x be the weight of a dwarf so that n^2x is the weight of a giant (n is a positive integer). So

$$D_ax + G_an^2x = D_cx + G_cn^2x \qquad (9)$$

Using (3) we have

$$G_c - G_a = (A - C)/(n^2 - 1) \qquad (10)$$

From (6) and (8) either

$$G_c - G_a = 4/(n^2 - 1), \, C = 3 \qquad (11)$$

or

$$G_c - G_a = 8/(n^2 - 1), \, C = 6 \qquad (12)$$

We require that $G_c - G_a$ is a positive integer so that only $n=3$ is a solution.

$$G_c - G_a = 1, C = 6 \tag{13}$$

From (3) and (6) we have

$$D_a + G_a = 14 \tag{14}$$
$$D_c + G_c = 6 \tag{15}$$

and (13) and (15) give

$$D_c + G_a = 5 \tag{16}$$

From (14) and (16) we have

$$D_a - D_c = 9 \tag{17}$$

However, from (1) and (17)

$$D_a = 10, D_c = 1$$

So (14) and (15) give

$$G_a = 4, G_c = 5$$

Now (3) and (6) produce

$$D_b + G_b = 10$$

Combining with (2) gives

$$G_b = 6, D_b = 4$$

In the same boat

The enemy agents are C and H.

The problem amounts to asking: which two crew members in the three consecutive sets of three position groups can reproduce the three corresponding boat inclinations that framed the group leader? We first need to find out who the group leader is, together with his qualitative relative weight (heavier or lighter than chosen weight) by finding the three position groups after the first set of interchanges, the initial and final sets of three groups being given. Information is provided by an analysis of the way the two tests must cooperate in order to identify just one imposter.

Consider 12 men in three position groups of four men all with equal weight X except the imposter (group leader) who has a weight Y that is either heavier or lighter than X. The boat may have three possible inclinations: (i) forward tilt, (ii) balance, (iii) backward tilt. Whichever of these the boat has, there are only three effects of rearranging the crew: unchanged inclination, or a change to one of the other two. The group leader is necessarily determined at the end of the second test if there are no more than three suspects presented to the second test, so that each can be linked to a different effect by appropriate placement.

The first test must decide between groups of suspects in order to present one suspect group to the second test. Here there can be no more than three groups so that each group can be linked to a different effect. With an initial forward tilt, there are eight suspects ABCDIJKL (all from the end position groups) which can only be divided into groups of 2,3,3.

I. First test

The rearrangement, consisting of four interchanges involving ABCEFGIL, must allow discrimination between the three groups of suspects by linking each group to a different effect.

(a) *Reversal group (two suspects)*
If the group leader were in this group, he would reverse the forward inclination to a backward inclination when the group is moved. Here only suspects participate in the interchanges, being moved from one end of the boat to the other. Since only one interchange can occur then the group has only two members.

(b) *Balance group (three suspects)*
If he were here, the forward inclination would become balanced when the group is moved. Suspects are moved from the ends to the middle, exchanging positions in three interchanges with three non-suspects in the middle.

(c) *Unchanged group (three suspects)*
All group members are suspects and none are moved. If the group leader were amongst them, the forward tilt would be left unchanged.

These suspect groups take care of the four interchanges and associate each group with a different effect of rearrangement.

2. Second test

A rearrangement in the second test permits discrimination between two or three suspects. Here the group (reversal, balance, unchanged) containing the group leader has already been identified after the first interchanges, and we must now identify the group leader in this group.

(a) *Two suspects*

Only the reversal group applies here. One suspect is placed in the middle (Y is identified by a resulting balance) and the other is placed at one end (Y identified by imbalance). Reference to the initial forward tilt positions gives the qualitative relative weight.

(b) *Three suspects*

These arise from the balance or unchanged groups. One suspect is placed at each end and one in the middle. For the special case where two of the three suspects were initially in the same position in the forward tilt distribution (e.g. two at back) then these are separated to each end with the third in the middle, in the second test. Reference to the initial distribution identifies Y and the qualitative relative weight.

Deductions

Test	Back	Middle	Front	Inclination
(0)	ABCD	EFGH	IJKL	Forward
(1)	?	?	?	?
(2)	ACHL	DIJK	BEFG	?

Initial suspects: ABCDIJKL
Four interchanges (0)–(1): ABCEFGIL

3. Unchanged group

The suspects not interchanged from (0)–(1) form the unchanged group $D_u J_u K_u$; note that H is not a suspect.

(1) D_u H $J_u K_u$

Since this group is not distributed from (1)–(2) according to specifications (2b), they are not being tested in the second test and the group leader cannot be amongst them.

4. Reversal groups

The possible reversal groups (see 1a) are those suspects that might exchange ends from (0)–(1), namely, AI,BI,CI,AL,BL,CL. The only possibilities that could be tested in (2) are AI,BI,CI (see 2a). It is given that the group leader occupies the same position in (1) and (2). After interchange of ends from (0)–(1), only B in BI satisfies this condition, so it is possible that the group leader is B and the reversal group participates in the second test. However, we can show that BI, too, is not presented to the second test.

Suppose BI were the reversal group so that

(1) D_uI_r H $B_rJ_uK_u$

Then with the reversal and unchanged groups determined, only ACL can be the balance group (see 1b) and must go to the middle from (0)–(1).

(1) D_uI_r $A_bC_bL_bH$ $B_rJ_uK_u$

However, it is given that A and G occupy the same position in (1). This violates the requirement of four in each position hence BI is not a reversal group and no reversal groups are tested in (2). Only a balance group can now appear in the second test.

5. Balance group

The possible balance groups (see 1b) – with the unchanged group already determined – are ABI,ABL,ACI,ACL,BCI,BCL. Only the possibilities ABI and BCI could be tested in (2) – see (2b). This gives BI in the tested balance group plus one of A or C, all three going to the middle in (1) – see (1b). So L and the remaining one of A or C are in the reversal group and exchange ends from (0)–(1).

(1) D_uL_r $(A/C)_bB_bI_bH$ $(C/A)_rJ_uK_u$

6. Final deductions

It is given that A and G occupy the same position in (1). With four in each group they must both be at the front, C must be in the middle and hence EF at the back.

(1) D_uEFL_r $B_bC_bI_bH$ $A_rGJ_uK_u$

The group leader must be I, being the only one in the balance group occupying the same position in (1) and (2). Hence I is heavier than X from (0) and (1),(2) are balanced.

Test	Back	Middle	Front	Inclination
(0)	ABCD	EFGH	IJKL	Forward
(1)	DEFL	BCHI	AGJK	Balance
(2)	ACHL	DIJK	BEFG	Balance

What two crew reproduce the three inclinations such that one is heavier than X by as much as the other is lighter (sum of their weights is $2X$)? Only the pairs CH,JK,EF are together in both (1) and (2) to give a balance. In (0), only C (lighter) and H (heavier) can produce the forward tilt.

Lost in space

In descending order, the ranking is B, D, C, A, E; the first two lied.

With the particular questions asked, a completely determined order only arises from having exactly one 'Yes' or one 'No' in five truthful answers.

So the two wrong answers must be a 'No', 'No' or a 'Yes', 'No'. Only inverting each of the last pair keeps exactly one 'Yes'. So inverting the reported 'Yes' either C or D lies. If it's C, only A as the other liar gives exactly two inversions. If it's D, then similarly, B is the other.

For these two cases, NNNNY and NYNNN are the deduced correct answers. Only E is trustworthy, being in both possibilities, so his correctly reported 'No' reduces us to the second case. The solution follows.

Pet hate

2	Angus	Tipple	Elephant	Nasher
4	Judy	McTumble	Cobra	Grumpy
6	Celia	Grout	Rhinoceros	Bubbles
8	Herbie	Blip	Tarantula	Stomper

Angus will not fit at number 8 and cannot be at 4 because then number 6 has no possible pet. If Angus is at 2, Herbie cannot be at 6 (one or two items would be correct in the first name column); and if Angus is at 6, Herbie cannot be at 2 (two or four items would be correct in the first name column). So either Angus is at 2 with Herbie at 8, or Angus is at 6 with Herbie at 4.

In both cases, the tarantula and elephant cannot be at 2 and 6, respectively, so only one of the cobra and rhinoceros is correctly positioned. This eliminates the second case for Angus because neither are possible.

For the first case for Angus, if the rhinoceros is at 8, the elephant must be at 2 and the cobra at 4 (two items correctly positioned); so the cobra is correct at 4, with the elephant at 2, the rhinoceros at 6, and the tarantula at 8.

The other three columns are solved by realising that two positions are known in each, neither is positioned correctly so a correct position occurs in exactly one of the other two places.

The three prisoners

Cell 4 had no prisoner and only statement (b) is false.

Did you fall into the trap of assuming that only the three prisoners are in the cells? In this case, one prisoner must make two statements, and due to the possible positions of the perceivers, he must make (a) and (c) or (b) and (d).

The first pair are contradictory, and for the latter pair there is no consistency in any of the three true/false combinations (involving the four statements) yielding the state of occupancy results.

The correct approach is to realise that since no two people are in the same cell when the statements are made, the warder must be visiting the vacant cell. By his own demand, he must make a statement, and since he is blind, it must be (b), (c) or (d). We must then consider the true/false combinations for the remaining three statements in each case. Only for the warder making (c) do we get consistencies, bearing in mind that he must be in 1 or 4 to make this statement. However, he cannot be in 1, since that was his first visit. So he is in 4 with (a) true, (b) false and (d) true.

The balanced bridge

Since the dog owner starts and finishes on the same bank, the two pairs of men, having exchanged sides using a balance, must both have an equal weight. Hence the dog owner has weight 1, 3 or 5.

The two lever operators cannot have total weight 1, and with weight 5 the dog owner must cross with two from the south bank leaving no lever operator. This means the dog owner has weight 3.

Two possible combinations remain for the men on the two banks. However, to have the first lever operator lighter than the second, we must have 2, 3, 4 on the north bank and 1, 5 on the south bank. Then 2, 3 exchange sides with 5 (lever operator is 1) followed by 4 with 1, 3 (lever operator is 2).

The broken pentomino

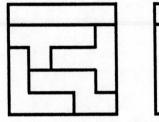 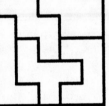

There is literally a twist in this puzzle because the new pentomino must be turned over to produce the solution!

One for the road

The strengths of the three men were 3, 5 and 9.

Let the three different pushing strengths be A,B,C so that

$$A > B > C > 0 \qquad (1)$$

The five men consist of two pairs of twins and the strongest man, that is, A, $2B$, $2C$. Two given conditions are

$$2C > B \qquad (2)$$
$$2B > A \qquad (3)$$

We can now list the 10 ways of arranging the three remaining men (with A always positive).

(a)	$A+B+C$	(f)	$A-B+C$
(b)	$A+B$	(g)	$A-C$
(c)	$A+C$	(h)	$A-B$
(d)	$A+B-C$	(i)	$B-C$
(e)	$B+C$	(j)	$A-B-C$

Since these must all produce a positive resultant push, we must have from (j) that

$$A > B + C \qquad (4)$$

Our aim now is to place these arrangements in order according to their resultant strengths using (1), (2), (3), and (4). For example, we can establish from (1) that (a) > (b) to (j) and that (b) > (c) to (j). The third place in the order is established by realising that, from (2),

$$A + C > A + B - C$$

and from (1)

$$A + C > B + C$$

and (c) > (f) to (j). The next in the order requires the use of (1) and (4). Continuing in this way we find that the descending order is

(i)	$A + B + C$	(vi)	$A - B + C$
(ii)	$A + B$	(vii)	$A - C$
(iii)	$A + C$	(viii)	$A - B$
(iv)	$A + B - C$	(ix)	$B - C$
(v)	$B + C$	(x)	$A - B - C$

The most effective arrangement where a man pushed at the front is $A + B - C$. We can now prove the following descending order from (1), (2), (3) and (4).

(i)	$A + B - C$	(vii)	B
(ii)	$2B$	(viii)	$A - B$
(iii)	A	(ix)	C
(iv)	$B + C$	(x)	$B - C$
(v)	$A - B + C$	(xi)	$A - B - C$
(vi)	$A - C$	(xii)	0

Since $A + B - C$ is 11 units and the order above represents different whole numbers, they form a descending sequence 11 to 0 and the solution follows.

Round the clock

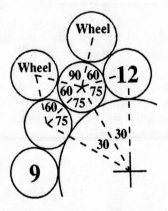

The wheel rotates six times each hour. Let the circumference of each small plate be C. As shown above, the wheel runs over $C/4$ of each plate. There are two additive components of the wheel rotation to consider as the wheel runs over each plate. If the circumference of a plate were stretched out into a linear path, since the radii of wheel and plate are equal, the wheel rotates $C/4$ due to the path length. However, the tangent to the plate at the point of wheel contact rotates through 90 degrees as the wheel rolls over the plate. This rotates the wheel an extra $C/4$ for each plate. So the total wheel rotation is $12(C/4 + C/4) = 6C$ with respect to the clock centre.

Cryptic cave lines

The rows are ODDS, MAIL, ACRE, NEED reading from top to bottom.

The keyword is DIRE which, since all its letters are different, can only be in a column or one of the two middle rows. It only fits the third column, since with the other possible positions and the given equations, the grid cannot be sensibly filled in.

Now from the three equations, the solid square is at least twelfth in the alphabet and must be the first letter of ?EED. Only N gives a sensible combination leading to the rows given above and the columns OMAN, DACE, DIRE, SLED.

Elixir of life

The old man is 135. In fact there are only two non-zero positive integer solutions in *A*, *B* and *C* to the equation:

$$100A + 10B + C = ABC(A + B + C)$$

which are $A = 1$, $B = 3$, $C = 5$ (different digits) and $A = 1$, $B = 4$, $C = 4$ (not different). The quick method is to interpret the cryptic message as 'The digits can be found by deleting M'. The two names in the puzzle, Neo-Hermet and Menpet, both have M and when deleted we have 'Neo-Heret' and 'enpet'. The first is an anagram of 'one-three' and the second is an anagram of the Greek word 'pente' meaning 'five'. If you think that's hard, try solving the equation (without a computer)!

The engineer's dilemma

Arnold must choose $N = 18$. Let A rotate about B at *x* revolutions per second (rps) and let the large wheel diameter be *y* times that of the small wheel. Then the small wheel rotates about its centre at yN (rps) and the centre of A (hence the circumference of A) rotates about B at $-x(1+y)$ (rps). Since we require that the circumference of A does not rotate we have

$$yN - x(1+y) = 0$$
$$N = x(1+y)/y$$

Since $x = 16$ and $y = 8$, we have $N = 18$.

Sum secret

The three numbers are 7868753, 8656865, and 9999988.
The sum can be broken down into three addition sums,

6 8 9 7	4 9 6 8	4 8 9 7
+ 2 9 6 8	+ 4 8 9 7	+ 3 8 5 6
9 8 6 5	9 8 6 5	8 7 5 3

```
    689              496              489
  + 296            + 489            + 386
  -----            -----            -----
    985              985              875


     69               49               49
   + 29             + 49             + 38
   ----             ----             ----
     98               98               87


      6                4                4
    + 2              + 4              + 3
    ---              ---              ---
      8                8                7
```

Taking a bath

Dan in Clancy's bath, Ben in Dan's, Clancy in Abe's, and Abe in Ben's.

Let Clancy have volume V. Then Dan, Abe, and Ben have volumes $2V$, $3V$, and $4V$, respectively. Let Ben have a volume V_0 of water in his bath initially. Then each bath has capacity $V_0 + 4V$, and Abe, Dan and Clancy have $V_0 + V$, $V_0 + 2V$ and $V_0 + 3V$ in their baths, respectively.

There are nine ways the brothers can all climb into incorrect baths. For each case, the water left in each bath after overflow can be calculated. Now the given condition 'The owner examined the baths . . .' requires that V_0 is smaller than V but greater than 0. Only one of the nine cases satisfies the final condition, the rest requiring V_0 less than or equal to 0 or V_0 greater than or equal to V. So $V_0 = V/2$ with $3V$ being the total additional water required. The solution follows.

The great escape

The prisoner clears the final door 13 minutes 25 seconds before the guard returns.

All five doors have a cycle which is a multiple of 35 seconds. Let this be one time unit. The first door opens every 3 units, the second every 2 units, the third every 5 units, the fourth every 4 units, and the fifth every 1 unit. If the guard leaves at 0 units' time then he will return when the doors next open simultaneously at 60 units, the lowest common multiple of 3,2,5,4,1. The maximum number of time units allowed between passing through the first and fifth doors must be 4 units (2 minutes 20 seconds) since 5 units (2 minutes 55 seconds) will sound the alarm. The prisoner cannot benefit from having two consecutive doors open at once since he has insufficient time to run through both. The doors must therefore open in order at 1 unit intervals. The only five successive times possible between 0 and 60 units are 33,34,35,36,37 which are respectively multiples of 3,2,5,4,1.

The prisoner must therefore wait 33 units after the guard leaves before entering the corridor whereupon all doors open successively. He clears the final door at 37 units, 23 units before the guard returns.

Selfish sons

The brothers who lie and therefore have money are the first, second and fifth.

Whoever has money lies about his own wealth, so the first son would not admit having money whether he had some or not. The fifth son therefore lies (5_L) and the third must have no money and tells the truth (3_T). The statements then made become as follows:

    (a)   1 reports 3 saying 'precisely one of my four . . .'
    (b)   2 reports 5 saying 'exactly two of my four . . .'
    (c)   4 says 'precisely three of my four . . .'
    (d)   4 reports 2 saying 'all of my four . . .'

In (c) we have either 4_L or 4_T.
(1) Assume 4_L then:

    from (c): one, two, or four of 1,2,3,5 have money.
           Hence $(1,2)_{TT}3_T5_L$ or $(1,2)_{LT}3_T5_L$;

    from (a): with 3_T and the assumption 4_L, it is not true that only one
           of $(1,2)_{TT}4_L5_L$ or one of $(1,2)_{LT}4_L5_L$ have money. So with 3_T
           we must have 1_L and hence we have 2T;

    from (b): 5_L sees $1_L2_T3_T4_L$ so cannot say 'exactly two of my four . . .'
           because he would not be lying as required. Statement (b)
           must therefore have 2_L which gives an inconsistency.

(2) Assume 4_T then:

from (c): three of 1,2,3,5 have money. Hence $1_L2_L3_T5_L$;

from (b): 5 sees $1_L2_L3_T4_T$ which is consistent with 2_L5_L;

from (d): 2 sees $1_L3_T4_T5_L$ which is consistent with 4_T2_L;

from (a): 3 sees $1_L2_L4_T5_L$ which is consistent with 1_L3_T;

The solution follows.

Wire wizards

The groups are BE (terminal group), FGH (terminal group) and ACD (non-terminal group).

If Glasgow is to identify three groups of wires consisting of X wires at one terminal, Y at the other, and Z free, we must have $X \neq Y$. Then the circuit tests in Glasgow show that

X wires make a circuit with Y others

Y wires make a circuit with X others

Z wires make a circuit with no others

With the right values of X,Y,Z formed in London, and the right connections into three pairs formed in Glasgow, then given the members of each group, London should be able to identify all the ends there through circuit testing all pairs separately. The only way this can be done is to make use of the identity of the groups X,Y,Z when making the connections in Glasgow. There are three possibilities as follows:

(a) one from X is paired with one from Y;

(b) one from Y is paired with one from Z;

(c) one from Z is paired with one from X.

The characteristic that a wire in a known group makes a circuit with a wire in a different known group will, for London, identify the wire, providing that the possibility arises once only. The three possibilities therefore allow two wires in each group to be identified in London.

This accounts for six of the wires, leaving two. If there are to be at least two in each group and $X \neq Y$ then the only feasible distribution is $X=2$, $Y=3$ and $Z=3$, if we define X as the terminal group with the lower number. (It is not necessary to identify the positive and negative terminals; only to distinguish between the terminals by having $X \neq Y$.) We then have one free wire in each of Y and Z which are identifiable by the characteristic that one wire in each of Y and Z makes a circuit with no other in London. So given the members of each group and the connections between the groups, London can identify all the ends.

However, we are presented with the situation in London that the members of the group are not known. From the given connections we only know that:

(1)   A is in a different group to G
       B is in a different group to C
       E is in a different group to F.
       So D and H are therefore free, but in different groups, one of
       them being a terminal group and the other being the non-terminal
       group Z.
       From the other information we know that:

(2)   Since F can make a circuit in Glasgow it is not in Z. It must be in
       X or Y and since it makes no circuit with G, either G is in the
       same group as F, or G is in Z.

(3)   B makes a circuit with H means they are in different terminal
       groups and from (1), D must therefore be in Z.

(4)   The fact that there are least 6 wires that A cannot make a circuit
       with puts it in Z, for in X or Y the number it could not make a
       circuit with would be all of $Z = 3$ and the rest of its own group
       (i.e. 1 or 2 remaining), a maximum of 5.

We now reason as follows. From (3) we have

GROUP	X,Y	Y,X	Z
MEMBER	B	H	D

and from (4)

	B	H	AD

From (2), G cannot be in Z since from (1), A and G are in different groups. So G is in the same group as F; that is, both are together in the first or second group. However, F and G cannot be in the first group since we then have B,F,G together (all wires forming connections) which violates the condition tied up with (1) that there are only two of the connected six in each group.

Hence F,G are together in the second group.

B   FGH   AD

Since A and G already connect the second and third groups, F in the second must connect to the first with E – see (1).

BE   FGH   AD

The final connection between B and C, from (1), gives C in Z.

The mathematical garden

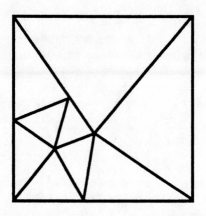

The arrangement subject to symmetrical variations is shown above.

The Three Bears

Daddy Bear ate 18 spoonfuls, Mummy Bear ate 36 spoonfuls, and Baby Bear ate 27 spoonfuls.

Let the amounts that these bears originally had in their bowls be *D*, *M* and *B* respectively, and their total consumption be *d*, *m* and *b* respectively. Then we can form the following three equations:

$$b = (1/2)(M - (1/3)d)$$
$$m = (1/2)(D - d)$$
$$d = B - (1/3)m$$

Putting d from the third equation into the second gives m, which then yields d from the third. Then substituting d into the first gives b.

Logical legacy

The red boxes totalled $40 000, the blue boxes totalled $35 000, and the green boxes totalled $25 000.

Suppose the blue statement is true and the others false. Then we have $10 000 in a blue and a red. Taking into account the false statements we have

	B (T)	G	R
(a)	10 + 15	(15 + 25)/(25 + 25)	(10 + 25)/(10 + 15)
(b)	10 + 25	15 + 15	10 + 25

The blue cannot possibly hold the highest total of the three. Assuming the green is true, then a green and blue each have $25 000, so

	B	G (T)	R
(a)	(25 + 15)/(25 + 10)	25 + 10	(10 + 15)/(15 + 15)
(b)	25 + 15	25 + 15	10 + 10

Here, the green cannot be the highest of the three. Finally, supposing that the red statement is true, a red and green each have $15 000, so that

	B	G	R (T)
(a)	25 + 25	15 + 10	15 + 10
(b)	(10 + 25)/(10 + 10)	(15 + 10)/(15 + 25)	15 + 25

Here, the possibility arises for red to have the highest total of the three. The first part of case (b) applies.

References

Beckmann, P. *A History of Pi* (The Golem Press : 1971).

Boyer, C. and Merzbach, U.C. *A History of Mathematics* (Wiley : 1989).

Bunt, L.N.H., Jones, P.S. and Bedient, J.D. *The Historical Roots of Elementary Mathematics* (Dover : 1988).

Clarke, B.R., *Test Your Puzzle Power* (Ward Lock : 1994).

Dudeney, H.A. *Amusements in Mathematics* (Dover : 1970).

Domoryad, A.P. *Mathematical Games and Pastimes*, p. 35 (Pergamon : 1964).

Fauvel, J. and Gray, J. (eds.) *The History of Mathematics: A Reader* (Macmillan Education : 1987).

Gardner, M. *Mathematics, Magic, and Mystery* (Dover : 1956).

Hadley, J. and Singmaster, D. 'Problems to Sharpen the Young' by Alcuin of York, in: *The Mathematical Gazette*, **76** (no. 475), pp.102–126 (1992).

Hordern, E. *Sliding Piece Puzzles* (Oxford University Press : 1986).

Li, Yăn and Du, Shírán (trans. Crossley, J.N. and Lun, A.W.-C.) *Chinese Mathematics* (Clarendon Press : 1987).

Loyd, S. Jr *Sam Loyd and his Puzzles – An Autobiographical Review* (Barse & Co, NY : 1928).

Newing, A. The Life and Work of H. E. Dudeney, *Mathematical Spectrum*, **2** (no. 2), pp.37–44, 1988–9.

Rubik, E. *et al.* (ed. Singmaster, D.) *Rubik's Cubic Compendium* (Oxford University Press : 1987).

Singmaster, D. *Sources in Recreational Mathematics, An Annotated Bibliography*, 6th preliminary edition, South Bank University, London, November 1993 (1993*a*).

Singmaster, D. Private correspondence, 17 July 1993, information originated from Will Shortz (1993*b*).

Stillwell, J. *Mathematics and Its History* (Springer-Verlag : 1989).

Wells, D. *Curious and Interesting Puzzles* (Penguin : 1992).

White, A. *Sam Loyd and his Chess Problems* (Leeds : 1914).

Printed in the United States
64841LVS00002B/71